The Power of Oneness

Solve Problems Synergistically Interconnecting All Possibilities

Kannappan Chettiar

Kannappan Chettiar

Copyright © 2025 Kannappan Chettiar

All rights reserved. No part of this publication may be reproduced, distributed or transmitted in any form or by any means, including photocopying, recording or other electronic or mechanical methods without the prior permission of the author, except in the case of brief quotations embodied in critical reviews and certain other non-commercial use permitted by copyright law. Permission requests, business or sales promotional use should be addressed to ceo@switchingbattery.com.

For licensing foreign or domestic rights, please email: ceo@switchingbattery.com

For sales enquiries and special bulk quantities, please contact our Order Services Department at sales@switchingbattery.com

First edition 2025
ISBN: 978-1-7378384-2-5 (Paperback)

Liberty of Congress
Control No: 2025900982

Published by:

Switching Battery Inc.
7180 Spumante Court
Gilroy CA 95020

Printed by:

Fizix Solar Innovations Pvt Ltd.
21, Sangam Rd, Lakshmipuram,
Chromepet, Chennai, Tamil Nadu 600044
Ph: + 91 8778223388

LIMITS OF LIABILITY / DISCLAIMER OF WARRANTY:
The author and publisher of this book have used their best efforts in preparing this material. While every attempt has been made to verify information provided in this book, neither the author nor the publisher assumes any responsibility for any errors, omissions or inaccuracies. The author and publisher make no representation or warranties with respect to the accuracy, applicability, fitness or completeness of the contents of this program. They disclaim any warranties (expressed or implied), merchantability or fitness for any purpose. The author and publisher shall in no event be held liable for any loss or other damages, including but not limited to special, incidental, consequential or other damages. As always, the advice of a competent professional should be sought.

Switching Battery© is a registered trademark of the author.

This book is for you— the marginalized, the overlooked, and the unrepresented. May it inspire hope, unity, and the power of oneness.

Contents

	Preface	i
1.	The Foundation of Oneness	1
2.	Electricity as a Metaphor for Life	17
3.	The Illusion of Separation	25
4.	The Logic of Coexistence	29
5.	Giving Up a Little to Gain Much More	33
6.	Energy as a Unifying Force	37
7.	The Economics of Giving	41
8.	The Balance of Progress and Preservation	45
9.	Timing as the Key for Interconnectedness	49
10.	Designing Systems for Oneness	55
11.	The Human Element in Interconnectedness	59
12.	Toward a Unified Future	63
13.	The Power of Hope and Action	67
14.	Beyond the False Dichotomy	71
15.	The Path to Collective Abundance	79
16.	Leadership for an Interconnected World	83
17.	The Responsibility of Knowledge	89
18.	The Beauty of Simplicity	95
19.	The Circle of Reciprocity	101
20.	The Future We Create Together	107
21.	Reflective and Proactive Awareness	113
22.	Beyond Binary: Open to Possibilities	119

Preface

The surest way to guarantee the extinction of the human race is to eradicate all plant life. Without plants, humanity—and countless other species that share this Earth—would eventually run out of breathable air. It's as simple, and as devastating, as that. This stark reality underscores a fundamental truth: humanity is not separate from nature but deeply intertwined with it. We are one part of a vast, interconnected equation.

The Power of Oneness is a call to understand and honor this profound connection. It urges us to see the world not as a fragmented collection of isolated systems but as an integrated whole, where every action sends ripples across ecosystems, societies, and future generations.

This book is not about lofty ideals or abstract philosophies; it is about practical, actionable strategies to align human systems with nature's timeless principles. These are the principles that have brought balance and sustained life on this planet for billions of years. By understanding our place within this intricate web of life, we can design technologies, policies, and lifestyles that reflect harmony over dominance and collaboration over conflict.

The time for change is now. The choices we make today will determine whether we contribute to the flourishing of life or its decline. Through this book, I present a new roadmap—a roadmap for living with purpose, building systems that support both humanity and the planet, and embracing the oneness that binds us all.

The message is simple yet profound: we are not separate from nature. We are nature. By recognizing this truth, we can forge a future that sustains and honors all life.

On a personal note, I humbly invite my family, friends, neighbors, coworkers, and collaborators to test the message and principle laid out in this book. These principles have guided me in solving problems that once seemed insurmountable, including innovations like the Switching Battery that solves the once unsolvable "energy trilemma"—a challenge that many top scientists, engineers and leaders deemed unsolvable because it required solutions that were simultaneously affordable, reliable, and sustainable. Yet, I believe I have achieved this with the creation of the world's first AC battery, delivering a 98% conversion efficiency compared to today's complex systems, which are less than 70% efficient, oversized, and prohibitively expensive.

My dream has always been to serve the bottom billion—the unrepresented population who face the greatest challenges in accessing clean and affordable energy. I deeply believe that the bottom of society forms a critical part of our global ecosystem. Helping the bottom billion is not merely an act of altruism; it is a logical choice—one that ultimately helps me, my family, and all of us. When we uplift the least advantaged, we strengthen the entire system, creating a more balanced and resilient world for everyone.

By working collectively, we can implement solutions that bring about zero carbon emissions immediately, reversing the devastation our current technologies have inflicted on the planet.

This book is an invitation to discover these principles for yourself and to validate their application in your own life and work. Nature's principles offer a path forward—a way to solve not just energy problems but many of the challenges we face. Together, we can co-create a future where humanity thrives in harmony with Mother Earth.

Kannappan Chettiar

> "A lamp can never light
> another lamp unless it
> continues to burn
> in its own flame."
>
> —*Rabindranath Tagore*
> *Nobel Prize in Literature 1913*

Chapter 1
The Foundation of Oneness

In this world, nothing exists in isolation. Everything is interconnected, and the sooner we truly internalize this truth, the better equipped we are to shape our lives and the world around us. What often appears separate—the sun, the oceans, and even individual human lives—is inextricably linked through a network of relationships that we seldom recognize. This is not merely an abstract idea; it is the foundation of reality itself. Grasping this interconnectedness has been the cornerstone of my work and my purpose.

My Unconventional Journey

My journey into energy systems and innovation has been anything but conventional. Unlike many in this field, I am neither an engineer nor a scientist by training. My academic background lies in law, where I hold four qualifications, including one from Berkeley Law School specializing in Energy and Clean Technology. I also hold degrees in Economics and Finance, Education and Learning, and have expertise in Change Management.

Teaching contract law has always been a passion of mine, but I find it counterproductive when so-called "contracts" expand into 30-page documents that no one truly reads or understands. I prefer simplicity—concise, clear agreements that people can engage with, rather than overly complex papers that serve little practical purpose. In essence, I believe in "contracting" contracts, distilling them to their essence.

This philosophy extends beyond law for me. I find the greatest satisfaction in reducing complexity to its simplest, most elegant form. Whether in contracts, energy systems, or life itself, I believe the world only truly makes sense when unnecessary complexity is stripped away, leaving clarity and functionality in its place.

My path took an unexpected turn into science, energy systems, and innovation—much like Michael Faraday, who, with no formal scientific education, made groundbreaking discoveries in electromagnetism through curiosity and relentless experimentation. Faraday's collaboration with James Clerk Maxwell, who provided the mathematical framework for Faraday's discoveries, highlights how unconventional thinkers can significantly contribute to science by bringing fresh perspectives and ideas.

Similarly, my pivot towards energy systems was not part of a predetermined plan but emerged out of necessity and curiosity. Leveraging principles from economics and combining with a legal analytical mindset, I began developing solutions that bridged gaps others had overlooked. For instance, I designed a simpler, more efficient battery management system that mimics the natural behavior of batteries. This system dynamically creates twin or analog voltages in real time and employs marginal charging methods, enabling solar energy to charge batteries even during low irradiance—a condition that traditionally hampers solar systems.

Reimagining Energy: A Journey Through History and Innovation

Unlike many experts in the field, I began my journey into energy systems by questioning historical assumptions and investigating the origins of the two key components of the Switching Battery: the battery itself and the switch. This exploration revealed not only the ingenuity of past inventors but also critical limitations that shaped my innovations.

1. The History of Batteries: Untapped Potential

My exploration started with the Persian battery, invented over 2,000 years ago. Often considered the world's first battery, it was not used for energy storage in the modern sense but for electroplating. While the Persian battery demonstrated potential, it lacked the necessary connections to function as an energy storage system. This realization underscored a foundational truth: batteries, much like pistons, have always been static components with potential energy. Pistons convert their potential into motion

3

through combustion, while traditional batteries rely on inefficient, unidirectional energy flows that limit their functionality.

Building on this, I discovered a critical limitation in traditional battery designs by studying Alessandro Volta, who connected individual batteries to create the first electric pile. While revolutionary in 1800's, Volta's approach remained limited to treating batteries as isolated units. Through observation, I uncovered a new possibility: instead of connecting batteries individually, it was more efficient to connect positive and negative nodes within the batteries themselves. This insight laid the groundwork for a patented Node Fusion hardware design that enables simultaneous charging and discharging, unlocking unprecedented flexibility and efficiency.

2. The Role of Switches: Intermittency Challenges

Next, I explored the history of the switch, credited to John Henry Holmes, who developed the modern electrical switch in 1884. While the switch allowed for precise control of electrical currents, it also introduced the problem of intermittency—the on-and-off interruptions that create inefficiencies in energy flow. These disruptions have hindered energy systems from achieving seamless operation, particularly in renewable energy systems.

The solution to this challenge came from an unexpected source: the traffic light, invented by Garrett Morgan in 1923. Morgan's introduction of the

4

amber light—a transitional phase between stop and go—reduced accidents and created a smoother flow of traffic. This inspired me to develop a similar transitional mechanism for energy systems. The absence of such a mechanism in traditional energy systems meant there was no effective way to bridge the gap between low-voltage and high-voltage states. Morgan's amber light provided the framework for solving this critical problem.

3. The Flaw of Zero: A Fresh Perspective

Perhaps the most profound realization came from revisiting the concept of zero, invented by Aryabhata over 1,500 years ago. Originally intended as a placeholder, zero was never designed to be a dynamic element in energy systems. In fact, transitioning between zero and one (on-off switching) in traditional binary systems is highly inefficient. It requires significantly more energy, similar to the energy required to "waking the dead".

This insight led me to develop an alternative approach: XY Logic, which eliminates the inefficiencies of binary switching. Instead of transitioning between zero and one, XY Logic enables energy to flow dynamically between parallel(X) low voltage and series(Y) high voltage configurations. The approach minimizes the distance energy must travel, allows energy systems to operate at lower frequencies, and increases overall efficiency.

5

4. The Switching Battery: A Holistic Energy System

The Switching Battery integrates these historical lessons into a unified, dynamic system that addresses the three core challenges of energy systems:

1. Batteries have historically been static components with potential energy, like pistons, but lacked mechanisms to convert that potential efficiently into motion or energy flow.

2. Switches introduced intermittency, creating disruptions in energy flow and hindering seamless operation.

3. The absence of a transitional mechanism to bridge low-voltage and high-voltage states left energy systems inefficient and incomplete.

The Switching Battery solves these challenges by combining batteries and switches into a single, holistic system. It mimics the bidirectional energy flow of pistons, but instead of relying on combustion, it uses solar irradiance as its power source. Through dynamic switching circuits, the Switching Battery enables energy to flow in parallel(X) and series(Y) configurations, creating a sinusoidal waveform that mirrors piston movements. The switching process is controlled by simple timing algorithms, allowing the system to operate efficiently, sustainably, and seamlessly.

In essence, the Switching Battery takes the static potential energy of batteries and transforms it into dynamic, bidirectional energy flow, bridging the gap between AC and DC with an unprecedented 98% efficiency. My innovation reimagines the roles of batteries and switches, not as separate components but as integrated parts of a unified energy system that is safer, more efficient, and aligned with the demands of the electronic age.

Lessons from History: Reimagining the Future

Reflecting on these discoveries, I realized that the greatest breakthroughs often emerge from reimagining what we think we already know. The Persian battery demonstrated potential but lacked connections for modern energy storage. The switch provided control but introduced interruptions. The amber light resolved traffic collisions but also provided an anology for smoother energy transitions. Even zero, originally a placeholder, became the foundation of computing but revealed inefficiencies in energy systems.

The Switching Battery builds on these lessons, integrating XY Logic and node-level connections to create a dynamic energy system that addresses historical limitations while paving the way for future innovation. By transforming batteries into predominantly electronic systems—powered 70% by XY Logic—it allows for seamless parallel(X)-series(Y) switching, increasing efficiency and safety while reducing costs and environmental impact.

This journey has shown me that progress is not about discarding the past but about building on its foundations, questioning assumptions, and uncovering new possibilities. The Switching Battery stands as a testament to this philosophy, bridging the gaps left by history and paving the way for a more sustainable and connected future.

Unlike other traditional batteries that are static, the Switching Battery system is 70% software and 30% hardware where the hardware is based on Node Fusion Technology and the software is based on the XY Logic. This enables a synergistic energy solution that is groundbreaking as it solves the once unsolvable problem known as the energy trilemma as it simplifies the complexity of energy systems by making them sustainable, affordable and reliable.

Collaborative Approch to Problem Solving

This work did not happen in isolation nor overnight as I worked with a number of great people over a span of 20 years. My ideas and concepts were significantly expanded and translated into actionable plans through the collaboration of three brilliant but humble scientists: Dr. Sergio Rivera (Colombia), Dr. Stephen D. Horowitz (United States), and Dr. Ermanno Pinotti (Italy)—all experts in power engineering and electronics. Their expertise was instrumental in refining my concepts and creating detailed plans that transformed abstract ideas into viable technological solutions.

8

These plans were then brought to life by my dedicated team of fresh, inexperienced electronic engineers from Chennai, India. While these young engineers lacked years of industry experience, they made up for it with enthusiasm, creativity, and an eagerness to learn. Guided by the groundwork developed by my teacher-scientist-friends, my Chennai team implemented the technology step by step, translating theoretical concepts into practical, market-ready solutions. Together, we turned innovation into reality, creating systems that are now available for sale and making a tangible impact in the energy sector.

This collaborative approach—combining vision, expert guidance, and the fresh energy of new talent—has been the cornerstone of my journey. It underscores a powerful truth: breakthroughs are not the result of a single individual but of collective effort, curiosity, and the willingness to push beyond conventional boundaries.

Understanding Similarities First

This ability to step back and analyze systems holistically enabled me to approach energy in an unconventional way. I began by studying the commonalities among human belief systems, particularly religions. While rituals and practices vary, all faiths converge on a central idea: a higher purpose for our existence on this material planet called Earth. This insight allowed me to see parallels in seemingly disparate fields. Just as religions express the same fundamental truth in different ways, alternating current

(AC) and direct current (DC) are simply different expressions of electricity. AC flows in waves, while DC is steady, but both are manifestations of the same force. The similarities go deeper. Just as both pistons and batteries are inherently static, they both possess potential energy that can be harnessed for movement. Pistons move when ignited by fire, converting chemical energy into mechanical energy. As explained earlier, the Switching Battery derives its power from solar irradiance, converting the energy of sunlight into electrical potential. This dynamic process uses switching circuits within the batteries to enable bidirectional energy flow—low voltage in parallel(X) and high voltage in series(Y)—mimicking the piston's movements to produce a sinusoidal waveform.

By reimagining how solar energy is harvested, stored, and transformed, the Switching Battery bridges the gap between AC and DC. It demonstrates that these two forms of electricity are not opposites but complementary expressions of the same underlying force. Recognizing this truth became a foundational metaphor for my work, inspiring innovations that align technology with the natural rhythms of energy flow and sustainability.

Observing Nature's Teachings

The lessons didn't stop there. I learned from the Bhagavad Gita that activity is preferred to inactivity, an idea that resonated deeply with me and helped me connect with the Zeroth Law of Thermodynamics.

10

This law, which governs the interplay between heat and cold, became a powerful metaphor for the cosmos. Heat represents activity, while cold denotes inactivity—yet both are essential and interdependent forces in the universe.

I drew further inspiration from the Yin-Yang philosophy, which illustrates that everything is interconnected and requires balance. Heat and cold, activity and inactivity, AC and DC—all require a form of control to maintain harmony. This balance is increasingly missing in our world, leading to challenges like climate change and social inequality. Recognizing this imbalance, I felt a calling to create solutions that restore harmony, not just in energy systems, but in the way humanity interacts with nature.

Rethinking AC Electricity

Generating AC electricity traditionally involves the mechanical motion of pistons, driven by the combustion of fossil fuels. This process, inherently inefficient, converts only about 30% of the coal, oil, or natural gas into usable power, with a substantial 70% lost in the process. Similar inefficiencies plague renewable energy systems, where oversized inverters lead to energy losses of 30-60%, undermining the potential of clean energy.

The Switching Battery system changes this paradigm entirely. It mimics the bidirectional energy flow of pistons without the need for burning fossil fuels or mechanical movement. It's revolutionary

11

technology utilizes dynamic switching between parallel(X) low voltage and series(Y) high voltage configurations, enabling bidirectional battery voltage by moving energy up and down, similar to how static pistons move using timed ignitions.

By converting DC directly into AC through advanced XY Logic-based software algorithms and Node Fusion Technology, the system achieves an unprecedented 98% conversion efficiency. The innovation transforms how electricity is generated, stored, and delivered, offering a sustainable, compact, and efficient alternative to traditional systems. The approach is so unique and impactful that it is protected under US Patent 12,040,638.

A Journey of Discovery

The breakthrough moment came when I recognized an oversight in traditional battery systems: they interrupt the load while charging. This flaw was similar to the buffering problem in early video streaming, where interruptions hindered smooth playback. By addressing this issue, I developed the Switching Battery system, which harmonizes energy flow much like buffering technologies revolutionized video streaming. Through simple timing algorithms, the Switching Battery connects DC and AC seamlessly, integrating solar energy with storage in a way that feels as natural as breathing.

But this innovation wasn't just technical—it was personal. After experiencing a heart attack, I

12

began to understand the interconnectedness within my own body. My heart performed the function of a switch, while my lungs worked similar to batteries, and my brain played the role of a logic controller that worked with my body's circuitry, the nervous system to ensure that all were operating in perfect harmony.

The realization of the hidden electrical world within the human body inspired me to create an intelligent battery system—one that mirrors nature's principles and fosters collaboration between hardware and software.

As I underwent the journey of an inventor, I was also forced to confront my ego—I met numerous experts who often dismissed my "great" ideas as impossible, bound by their reliance on conventional methods and an unwillingness to learn something different. I learned that ego separates us, erecting walls that block progress. By letting go of my ego and embracing *amor fati*—acceptance of fate—I allowed life to guide me, even in the face of resistance and doubt. My only goal was to inform them—and now you—that we are all connected and that understanding this connection is the key to unlocking true progress.

The Principle of Interconnectedness

What I have come to realize is simple yet transformative: everything is related. This understanding has reshaped my perspective, not only about technology but about life itself. The intercon-

13

nectedness of systems, people, and the planet forms the foundation of Oneness. Recognizing these connections is essential to innovation, problem-solving, and living in harmony with the world.

Consider electricity: many view alternating current (AC) and direct current (DC) as opposites. AC flows in waves, while DC is steady. But at their core, both are simply forms of electricity, delivered differently. I often compare DC to a square and AC to an ellipse—different in appearance, yet the same in essence. This is not just about electricity; it's a metaphor for life. We may seem different in form or function, but we are all part of the same system.

Nature demonstrates this principle everywhere. Water cycles from the sky to the earth, into rivers, and back to the ocean. Plants and humans exchange gases in a perfect balance—oxygen for us, carbon dioxide for them. Strangely, humanity often acts as though it is separate, harming the environment without realizing that this harm ultimately returns to us. It's like punching your left hand with your right—it makes no sense.

The Switching Battery system embodies this principle of interconnectedness. It harmonizes AC and DC electricity, integrates solar energy with storage, and minimizes waste. By capturing energy even in low-light conditions and maximizing output throughout the day, it aligns with natural systems rather than working against them. More importantly, it makes clean energy affordable and accessible,

14

empowering those who are often excluded from traditional systems due to cost.

A Logical and Ethical Choice

Helping the bottom billion—those most marginalized by traditional systems—is not just ethical; it's logical. The bottom of society forms an essential part of our global ecosystem. By uplifting them, we strengthen the entire system, creating balance and resilience that benefit everyone, including ourselves. Sharing resources does not create scarcity; it creates abundance. This is nature's way, and it is the way forward for humanity.

Take something as simple as breathing. Every breath connects us to plants and the planet, demonstrating our dependence on each other. Ignoring this connection is not just unwise—it's dangerous. Living in harmony with these systems isn't optional; it's essential.

This book is about these truths. It offers not just a theory but a practical framework for solving problems synergistically. By understanding how systems work together, we can create solutions that are efficient, sustainable, and accessible. We can build a future where humanity thrives, not in opposition to nature, but as a part of it.

We are all one, and when we act like it, life becomes not just sustainable but beautiful.

Chapter 2
Electricity as a Metaphor for Life

Electricity is often seen as a technical concept, confined to wires, circuits, and devices. But when we look closely, it reveals profound lessons about life itself. AC (alternating current) and DC (direct current) are more than just forms of electricity; they are metaphors for how systems, relationships, and even people operate. Understanding them helps us see the interconnectedness of everything and provides a framework for thinking about progress, harmony, and balance.

The Interconnection of Time and Electricity:

At their core, AC and DC are deeply intertwined with the concept of time. DC represents constancy—a steady, unchanging flow of energy, much like the timeless stability found in enduring principles or foundational truths. AC, on the other hand, introduces the dimension of periodicity—a rhythmic alternation that mirrors the natural cycles of life, such as day and night, the seasons, or the ebb and flow of human emotions.

Mathematically, DC can be visualized as a square: stable, with defined boundaries and a uniform voltage over time. AC, by contrast, resembles an ellipse: dynamic, oscillating between positive and negative values in smooth, continuous waves. Despite their apparent differences, these forms are not opposites but complementary facets of the same phenomenon. Their mathematical and functional conversion is straightforward when we recognize that DC is simply AC with an infinite wavelength, and AC can be derived from DC by introducing periodic variation.

How DC Mimics AC Using the Principle of Duality: The principle of duality means we work with 4 nodes: two positive and two negative. These nodes enable us to dynamically switch between parallel(X) and series(Y) configurations, achieving behavior that mimics AC.

Here's how it works:

1. **Parallel and Series Switching:**

   - In parallel, the voltage remains constant, representing a stable, lower-voltage state.

   - In series, the voltage adds up, representing a higher- voltage state.

   - By switching between parallel and series, the system dynamically moves voltage up and down, much like the bi-directional flow of AC. This is analogous to the motion of pistons, where the upward and downward strokes create alternating energy.

2. **Generating Positive, Negative, and Neutral States:**

   - **Positive Zone:** Achieved by combining series and parallel configurations to increase voltage, mimicking the positive peak of AC.

   - **Neutral Zone:** Obtained through zero-parallel switching, where the system stabilizes at a midpoint voltage. This acts as the baseline or "zero" point in the AC waveform.

   - **Negative Zone:** Achieved using an H-bridge method, which reverses the polarity of the voltage. This mirrors the negative peak of AC, completing the waveform.

3. **Mimicking the AC Waveform:**

   - By timing switches between parallel (low voltage) and series (high voltage) configurations, while using the H- bridge for polarity inversion, the DC system generates a seamless alternating pattern resembling the AC waveform.

   - This pattern directly mimics the sine wave of AC, creating bi-lateral movement from a purely DC source.

The result is a DC system that seamlessly replicates AC behavior by achieving:

- The positive peak by combining parallel and series.

- The neutral state through zero-parallel switching.

- The negative peak via polarity inversion using an H- bridge method.

This method eliminates the need for traditional AC generators and converters, enabling DC systems to operate as a direct substitute for AC, bringing simplicity, efficiency, and versatility to modern energy systems.

Beyond the Surface Differences:

DC flows in a single direction, providing a constant energy source, while AC shifts directions periodically, rising and falling in a sinusoidal wave. These two forms, though different in operation, are complementary methods of delivering electricity.

This duality extends beyond electricity. In life, some things are constant, like DC, providing stability and dependability. Others, like AC, operate in cycles, bringing rhythm and variation. Neither is inherently better; both are necessary. DC powers small, precise systems like batteries and electronics, while AC efficiently delivers power over long distances. Together, they meet different needs, yet their essence is the same.

Lessons from Nature and Systems:

Nature inherently embraces this balance—cycles of day and night, changing seasons, and our own breathing all reflect an alternation between activity and rest. Aligning with these rhythms allows life to flow more effectively, much like using AC and DC appropriately in energy systems.

Take relationships as another example. Some connections provide steady, reliable support—the DC of our lives. Others bring excitement, challenge, and change—the AC. Both are essential. Just as a battery and a power grid serve different purposes, so too do the people and experiences in our lives. Recognizing this helps us value each for what it brings, rather than expecting everything to serve the same function.

The Role of XY Logic:

In problem-solving, we often default to binary thinking: yes or no, right or wrong, one solution or another. But life is not binary; it is dynamic, like AC. Challenges are best addressed by recognizing their complexity and addressing them with flexibility and creativity. This is the essence of what I call XY Logic—the ability to move between extremes and find balance, much like alternating between AC and DC as needed.

In energy systems, XY Logic is about switching between parallel(X) and series(Y) connections to optimize performance. In life, it's about knowing when to push forward(X) and when to step back(Y), when to hold steady(X) and when to adapt(Y). It's about seeing the whole picture and understanding that different approaches are not conflicting but complementary.

Applications and Innovations:

This perspective has influenced my work profoundly. The Switching Battery AC System unites AC and DC, maximizing efficiency by combining the stability of DC with the adaptability of AC. This innovative integration exemplifies the principle of harmony —leveraging differences to create superior solutions.

The same principle applies to addressing global challenges like climate change, inequality, and resource scarcity. These problems cannot be solved with a single, unchanging solution. They require systems that adapt, innovate, and integrate multiple approaches. The Switching Battery AC System is one small example of this, combining the strengths of AC and DC to create a more efficient and sustainable way to deliver energy. It's a model for how we can approach other challenges, by embracing complexity and working with, rather than against, the natural flow of life.

A Unified Framework:

As we move forward, it's essential to remember that progress doesn't come from choosing one path over another but from finding harmony between them. AC and DC are not rivals; they are partners. The same is true for people, ideas, and systems.

War of Currents: Both Thomas Edison (with DC electricity) and Nikola Tesla (with AC electricity) only saw their respective difference resulting in the war of currents.that continues to destroy our tangible assets by burning fossil fuel. When we stop seeing differences as obstacles and start seeing them as opportunities for collaboration, we unlock the potential for something far greater.

Electricity reveals that everything is interconnected, with balance as the cornerstone of progress. The principles of AC and DC teach us to integrate diverse approaches, ensuring optimal outcomes in energy, relationships, and problem-solving. This understanding is not just theoretical; it is practical, and it is essential for creating a world that works for everyone.

Chapter 3
The Illusion of Separation

Humanity's greatest challenge is not technology or resources but perception. For centuries, we've operated under the illusion that we are separate—separate from nature, separate from one another, and separate from the systems that sustain us. This illusion has driven conflict, waste, and destruction, undermining the very foundation of life. To move forward, we must recognize that separation is not real; it is a construct of our limited perspective.

Consider the world's resources. We mine metals, extract oil, and chop down forests, treating these actions as isolated events. But every resource we take disrupts a larger system. Mining scars the earth and pollutes water. Burning oil fills the air with carbon. Deforestation destroys ecosystems that regulate our climate. These actions are not separate; they are deeply interconnected. When we harm the planet, we harm ourselves, even if the consequences are delayed or indirect.

This illusion of separation extends to our relationships with each other. Countries go to war over land, religion, or ideology, as if their interests are

25

distinct. Corporations exploit workers and the environment for profit, as if their success exists in a vacuum. Individuals hoard wealth and resources, as if their gains come at no cost to others. But none of this is sustainable. The pain we inflict on others— whether people or the planet—will inevitably circle back to us.

The illusion persists because it's convenient. It's easier to see the world in fragments than to grapple with its complexity. It's easier to believe we can take without consequence than to acknowledge our responsibility. But the cost of this convenience is staggering. Climate change, inequality, and resource depletion are not isolated problems; they are symptoms of a worldview that denies interconnectedness.

So how do we break free from this illusion? The first step is awareness. We must actively look for the connections we've been conditioned to ignore. Technology can help us do this. Satellite imagery reveals the shrinking rainforests and melting ice caps, showing us the global impact of local actions. Data analytics trace the flow of resources, money, and energy, revealing hidden patterns of interdependence. Science uncovers the complex ecosystems that sustain life, reminding us that we are not above nature but a part of it. But awareness is not enough. We must act on what we know. This starts with shifting our mindset. Instead of asking, "What can I take?" we must ask, "What can I contribute?" Instead of competing for short-term gains, we must collaborate for long-term sustainability. This requires rethinking our systems—from energy and economics to governance and education—to reflect the reality of interconnectedness.

In my work, I've seen how embracing interconnectedness can transform systems. The Switching Battery AC System, for example, was designed to integrate seamlessly with solar panels, batteries, and electrical grids. It doesn't work in isolation; it works because it recognizes and leverages connections. By eliminating inefficiencies like inverters and grid losses, it shows how we can create solutions that are not only more effective but more aligned with the world's natural rhythms.

This same principle can apply to how we live. Imagine if businesses operated not just for profit but for mutual benefit, reinvesting in their workers and communities. Imagine if governments collaborated across borders to solve shared challenges like climate change and resource scarcity. Imagine if individuals saw their success not as a solo achievement but as part of a larger collective effort. These are not utopian ideals; they are practical necessities if we are to thrive.

At its core, breaking the illusion of separation requires humility. We must accept that we are not the center of the universe but a thread in its vast web. This is not a loss but a gain. When we recognize our place in the whole, we unlock the potential for collaboration, innovation, and abundance. We see that giving up a little—whether it's time, resources, or

27

control—can lead to much greater rewards for everyone.

The illusion of separation has held us back for too long. It's time to see the world as it truly is: interconnected, interdependent, and inseparable. When we do, we will not only solve the challenges we face but create a life that is richer, more sustainable, and more beautiful for all.

Chapter 4
The Logic of Coexistence

If everything is interconnected, then our success depends not on domination but on coexistence. Coexistence is not just a moral ideal; it is a logical necessity. Systems, whether natural or human-made, function best when they operate in balance, supporting each other instead of competing destructively. To achieve this, we must embrace a mindset of mutual benefit, giving up a little to gain much more.

Consider nature: Ecosystems are not battles for survival; they are networks of collaboration. Trees provide oxygen and absorb carbon dioxide, supporting animals and humans. In return, animals contribute to seed dispersal and soil enrichment. The balance is not accidental; it is essential. When one part of the system overreaches like when deforestation removes too many trees the entire ecosystem suffers. The lesson is clear: coexistence sustains life, while unchecked competition destroys it.

The same principle applies to humanity. History is full of examples where collaboration led to growth and conflict led to collapse. The European Union, for instance, was founded on the idea of

29

cooperation after centuries of war. By pooling resources and aligning policies, member nations have achieved a level of stability and prosperity that individual efforts could not have accomplished. On the other hand, global challenges like climate change have worsened because nations continue to act in isolation, prioritizing short-term gains over shared solutions.

At a smaller scale, this logic applies to personal relationships and communities. Families thrive when members support each other, sharing responsibilities and benefits. Businesses succeed when employees work together, leveraging diverse skills toward common goals. In contrast, selfishness and mistrust erode these bonds, leading to inefficiency and failure. The evidence is everywhere: coexistence is not just idealistic; it is practical. The challenge lies in implementing this logic consistently. Coexistence requires us to think beyond ourselves, to consider the needs of others and the long-term consequences of our actions. This is not easy in a world that often rewards individualism and competition. But the rewards of coexistence far outweigh the sacrifices. Giving up a little whether it's resources, time, or control creates the conditions for greater collective success.

This is the principle behind the Switching Battery AC System. It is designed not as a standalone solution but as a collaborative component within a larger energy ecosystem. By integrating with solar panels, batteries, and even small grids, it maximizes efficiency and minimizes waste. It shows how technology can embody coexistence, working with existing systems instead of against them. The result is not just better energy performance but a model for how interconnected solutions can drive progress.

This logic can be extended to address global challenges. Take resource distribution, for example. The world produces enough food to feed everyone, yet millions go hungry because of inequitable systems. A cooperative approach sharing surplus food, reducing waste, and improving distribution could eliminate hunger without requiring more resources. Similarly, in energy, developed nations could share renewable technologies with developing ones, accelerating the global transition to sustainability. These actions benefit everyone, creating stability and reducing the risks of conflict and environmental collapse.

The logic of coexistence also applies to innovation. Many breakthroughs happen not in isolation but through collaboration. The internet, for instance, emerged from interconnected efforts by scientists, governments, and private companies.

Open-source platforms continue this tradition, enabling people worldwide to contribute to and benefit from shared knowledge. Coexistence in innovation accelerates progress by pooling ideas, resources, and expertise.

However, coexistence is not passive. It requires active effort to build and maintain relation-

31

ships, systems, and policies that prioritize mutual benefit. This starts with a change in mindset were we must move away from zero-sum thinking that one's gain is another's loss and embrace win-win solutions. We must recognize that our success is tied to the success of others, and that cooperation is the most effective path forward.

In practical terms, this means making decisions that balance individual and collective needs. It means investing in systems that uplift the marginalized, because their well-being strengthens society as a whole. It means prioritizing sustainability, because a healthy planet is the foundation for all human activity. It means seeing every challenge as an opportunity to collaborate, not compete.

Coexistence is the logic of interconnectedness in action. It is the understanding that we are stronger together than apart. By embracing this logic, we can create systems that are not only efficient but also resilient, fair, and sustainable. Whether in energy, relationships, or global governance, coexistence is not just a better way; it is the only way forward.

Chapter 5
Giving Up a Little to Gain Much More

Human nature often pushes us to accumulate as much as we can—money, resources, power. But this instinct, while understandable, is shortsighted. True progress and abundance come not from taking as much as possible but from giving up a little for the greater good. This principle, though simple, is transformative when applied to our personal lives, societies, and global systems.

At first glance, giving something up may seem like a loss. But in reality, it is an investment. In nature, this principle is everywhere. Trees shed their leaves to conserve energy for the winter, ensuring survival and future growth. Rivers surrender their waters to the oceans, fueling cycles that bring rain and replenish the earth. These sacrifices are not losses; they are essential exchanges that sustain life.

For humans, the same logic applies. Take relationships as an example. Compromise is often seen as a weakness, but it is the foundation of strong partnerships. Giving up a little—whether it's time, ego, or resources—builds trust and cooperation, creating

33

bonds that deliver far greater rewards over time. In contrast, selfishness isolates us, reducing what we can achieve together.

This principle extends to economics. Companies that prioritize short-term profits over long-term relationships often find themselves struggling. On the other hand, businesses that invest in their employees, customers, and communities—even at a short-term cost—tend to thrive. The same is true for nations. Sharing resources and technology with less developed regions may seem like a sacrifice, but it creates stability, markets, and partnerships that benefit everyone.

In my work, this principle has been a guiding force. The Switching Battery AC System, for example, was designed with affordability and accessibility in mind. By keeping costs low and maximizing efficiency, it provides clean energy to those who need it most. This required giving up certain profit margins, but the result is a solution that can reach more people and create a larger impact. This is the power of giving up a little: it opens the door to something much greater.

At a global level, this principle is crucial for addressing the challenges we face. Climate change, resource scarcity, and inequality cannot be solved by isolated efforts or selfish policies. They require cooperation, where nations and individuals are willing to contribute to solutions that benefit everyone. This might mean cutting emissions, sharing renewable energy technologies, or investing in education and healthcare for underserved populations. These actions may involve sacrifices, but the alternative—a fractured, unstable world—is far more costly.

The logic of giving up a little is not just about altruism; it is about creating systems that work better for everyone. When we prioritize the collective good, we build trust, resilience, and opportunity. A society where people hoard resources breeds fear and conflict. A society where people share creates abundance and stability. This is not idealism; it is logic grounded in evidence.

Consider the story of renewable energy. Fossil fuels have dominated for centuries because they prioritize short-term gains—rapid extraction and profit—without accounting for long-term costs like environmental destruction and climate change. Transitioning to renewables requires upfront investments and sacrifices, but the payoff is immense: cleaner air, sustainable energy, and a stable climate. Giving up a little now secures a far greater future.

This principle also applies to personal choices. Imagine a person who devotes a portion of their time to mentoring others. At first, it may seem like they are giving up valuable hours they could use for themselves. But in the process, they build relationships, expand their influence, and create a ripple effect of positive change that enriches their own life as well as others. Similarly, when we reduce our consumption or switch to more sustainable habits, we may give up convenience, but we gain a healthier planet and a better quality of life for future generations.

Ultimately, giving up a little is not about loss; it is about unlocking potential. It is about recognizing that we are part of a larger system and that our individual choices have collective consequences. When we align our actions with this understanding, we create a world where everyone can thrive.

The power of this principle lies in its universality. It applies to energy systems, economic models, relationships, and even personal growth. Whether it's a tree shedding its leaves, a nation sharing its resources, or a person lending a helping hand, the act of giving up a little often leads to gains that are exponentially greater.

The question is not whether we will make sacrifices but how we will make them. Will we cling to short-term gains and ignore the bigger picture, or will we invest in a future where everyone benefits? The answer will define not just our success but our survival. By giving up a little, we can gain much more—for ourselves, for others, and for the planet.

Chapter 6
Energy as a Unifying Force

Energy is the foundation of life. It flows through everything we see and do, connecting people, systems, and the natural world. Without it, nothing moves, nothing grows, and nothing thrives. But energy is not just a physical necessity; it is also a powerful metaphor for the interconnectedness of all things. Understanding energy— how it works, how it flows, and how it can unite us—is essential to building a world that functions harmoniously.

From the smallest atom to the largest galaxy, energy binds the universe together. The sunlight that powers a plant's growth is the same energy that fuels our bodies when we eat. The wind that turns a turbine to generate electricity is driven by the sun's heat. Even the energy we expend in our daily lives is part of this grand cycle. Energy connects us to nature and to each other in ways we often take for granted.

For centuries, humanity has sought to harness energy for progress. We discovered fire, learned to farm, and built machines that transformed our world. But along the way, we began to see energy as something to dominate rather than something to collaborate with. Fossil fuels became the primary source of power, providing convenience and growth but at a devastating cost to the planet. This approach treats energy as a commodity, ignoring its role as a unifying force. Renewable energy offers a chance to restore this balance. Unlike fossil fuels, which are finite and destructive, renewables tap into the natural flows of the earth—the sun, wind, and water. These sources are infinite, sustainable, and inherently interconnected. By shifting to renewables, we align ourselves with the planet's rhythms rather than working against them. This is not just an environmental imperative; it is a philosophical one. To harness energy sustainably is to acknowledge our place within the larger system of life.

This philosophy is at the heart of the Switching Battery AC System. It is designed to work with renewable energy, maximizing its potential while minimizing waste. By capturing energy during low-light conditions and converting it efficiently, the system ensures that no sunlight goes to waste. It eliminates the need for inverters and transformers, reducing complexity and increasing efficiency. But more than that, it demonstrates how technology can integrate seamlessly into the natural flow of energy, bridging the gap between humanity and nature.

Energy is also a social force. Access to energy determines quality of life. It powers homes, schools, and hospitals. It enables communication, transportation, and industry. Yet, billions of people around the world still lack reliable access to electricity. This inequality is not just a technical problem; it is a moral one. If energy connects us, then denying it to some denies the interconnectedness of all.

The solution lies in rethinking how we distribute and use energy. Decentralized systems, like the Switching Battery AC System, provide a way to bring power to remote and underserved areas. By combining local solar generation with efficient storage, these systems create independence and resilience. They empower individuals and communities, reducing reliance on centralized grids and fossil fuels. This is energy as it should be: a force that unites and uplifts.

Energy also teaches us about balance. In physics, energy is never created or destroyed; it only changes form. This principle mirrors the cycles of life. What we take from the earth must eventually be returned. Over-extraction leads to imbalance, just as overconsumption leads to waste. The key is to find harmony—to use energy wisely, recycle it where possible, and ensure that its flows benefit all parts of the system.

This balance is reflected in the concept of XY Logic, which underpins the Switching Battery AC System. By dynamically switching between parallel and series configurations, the system adapts to changing conditions, optimizing performance without wasting resources. This logic can be applied to more than just energy systems. It is a way of thinking that emphasizes flexibility, adaptation, and integration—

qualities that are essential for solving complex problems in any field.

Ultimately, energy is a reminder of our interconnected-ness. It flows through every aspect of life, linking the physical, social, and philosophical. When we understand energy as a unifying force, we see that our choices about how to produce, distribute, and consume it are not just technical decisions but moral ones. They reflect our values and our vision for the future.

The path forward is clear. We must move away from systems that treat energy as a commodity to be exploited and embrace systems that respect its role as a connector and sustainer of life. By aligning our energy systems with the natural world and with each other, we can create a future that is not only sustainable but truly interconnected. This is the power of energy—to unite, to uplift, and to illuminate the path to a better world.

Chapter 7
The Economics of Giving

The modern world often measures success by accumulation: wealth, resources, and power. But true progress lies not in how much we keep, but in how much we share. The economics of giving—a principle rooted in both logic and compassion—demonstrates that collective well- being creates far greater rewards than individual hoarding. When we give, we invest in a system where everyone, including ourselves, benefits.

Economics is often framed as a zero-sum game: for one person to gain, another must lose. This perspective fuels competition and inequality, but it is fundamentally flawed. The real economy—the one that sustains life—is built on interdependence. Just as ecosystems thrive through cycles of exchange, human systems flourish when resources circulate freely and fairly. When people, businesses, and nations give to others, they don't lose; they create value that returns to them in unexpected and multiplied ways.

Take infrastructure as an example. Governments that invest in public goods like education, healthcare, and transportation often see significant returns. An educated workforce boosts innovation and productivity. Accessible healthcare reduces long-term costs and improves quality of life. Efficient transportation systems enable commerce and mobility. These investments benefit individuals and communities, creating a cycle of growth that lifts everyone.

The same principle applies to businesses. Companies that pay fair wages, invest in employee development, and prioritize sustainability often outperform those that focus solely on short-term profits. When workers are treated well, they are more productive and loyal. When customers see that a business values ethics, they respond with trust and support. Giving creates relationships, and relationships create value.

This logic extends to global challenges. Climate change, for example, cannot be solved by isolated efforts. Developed nations must invest in renewable energy and share these technologies with the rest of the world. This may seem like a sacrifice, but it builds a stable global economy and reduces the risks of environmental disaster that would harm everyone. Similarly, addressing inequality through fair trade, debt relief, and education creates markets and partnerships that benefit all nations. Giving at this scale is not charity; it is enlightened self-interest.

The Switching Battery AC System embodies this principle. By making clean energy affordable and accessible, it empowers communities to take control of their energy needs. This creates a ripple effect: individuals save money, businesses become more efficient, and entire regions gain independence from costly and polluting fossil fuels. The system's design reflects the economics of giving: it sacrifices complexity and profit margins to deliver a solution that benefits more people. In doing so, it creates a market that grows exponentially as trust and accessibility increase.

Giving also transforms personal relationships and communities. When people contribute their time, skills, or resources to help others, they build bonds of trust and reciprocity. These connections create a social fabric that supports everyone, especially in times of crisis.Communities that prioritize giving are more resilient, innovative, and joyful. They prove that generosity is not a cost but an asset.

The act of giving requires a shift in mindset. It challenges the fear of scarcity—the belief that there is not enough to go around. This fear drives hoarding, competition, and inequality. But when we recognize that resources are most effective when shared, we see that abundance is possible. The sun, for instance, provides enough energy to power the entire planet many times over. The challenge is not the scarcity of energy but the systems we've built to distribute it. By giving up outdated technologies and adopting renewable, decentralized systems, we can unlock this abundance.

43

The economics of giving is also about timing. A small investment at the right moment can prevent far greater costs later. Early action on climate change, for instance, avoids the catastrophic expenses of disaster response and recovery. Supporting education for underserved communities reduces poverty and crime, saving billions in social and economic costs. Giving, when done strategically, is not only compassionate but efficient.

Ultimately, the economics of giving reveals a profound truth: we are all connected. Our success depends on the success of others, and our well-being is tied to the systems we support. By embracing this principle, we move from a mindset of scarcity to one of abundance, from competition to collaboration, and from fear to hope.

The path forward is clear. We must design systems—in energy, economics, and society—that prioritize giving over taking, collaboration over isolation, and long-term well-being over short-term gain. This is not just a moral imperative; it is a practical strategy for creating a world where everyone can thrive. The more we give, the more we gain—not just as individuals but as a collective. This is the true economy of life.

Chapter 8
The Balance of Progress and Preservation

Progress is often celebrated as humanity's greatest achievement. We have built cities, cured diseases, and explored the depths of space. But progress comes with a cost. In our pursuit of advancement, we have depleted resources, destroyed ecosystems, and disrupted the delicate balance of the planet. The challenge now is clear: how do we continue to progress while preserving what sustains us? This balance is not just desirable; it is essential.

The conflict between progress and preservation arises from a fundamental misunderstanding. We often view them as opposing forces, assuming that to advance we must exploit, and to preserve we must stagnate. But this is a false dichotomy. True progress is not about taking from the planet but working with it. Preservation is not about resisting change but guiding it toward harmony. When we align these two forces, they become mutually reinforcing.

Nature provides the blueprint for this balance. Forests grow and thrive while maintaining the ecosystems they support. Rivers flow, nourishing life

45

without exhausting their source. These systems demonstrate that growth and sustainability are not opposites but partners. The key is efficiency and regeneration—using resources wisely and ensuring that what is taken is replenished.

Human systems, however, have often ignored this principle. Industrialization brought incredible advancements, but it was built on extraction without renewal. Fossil fuels, deforestation, and mass production prioritized short-term gains over long-term stability. The result is a world where progress has outpaced preservation, leading to climate change, resource depletion, and social inequality.

Rebalancing requires a shift in how we define success. Instead of measuring progress solely by economic growth or technological breakthroughs, we must also consider environmental health, social equity, and resilience. These factors are not barriers to progress; they are indicators of its quality. A system that advances at the expense of these factors is not progress at all but a step toward collapse.

This philosophy is embodied in the Switching Battery AC System. It represents progress in energy technology—a system that maximizes efficiency, eliminates waste, and integrates with renewable sources. At the same time, it prioritizes preservation by reducing reliance on fossil fuels and enabling decentralized, sustainable energy systems. It shows that innovation can enhance, rather than undermine, the balance of progress and preservation.

Balancing these forces also requires understanding scale. Global challenges like climate change demand large-scale solutions, but these solutions must be implemented locally. A wind farm or solar array can power an entire region, but its success depends on how it integrates with local communities, ecosystems, and needs. The Switching Battery's compact and adaptable design reflects this principle, proving that solutions can be both scalable and specific.

Timing is another crucial factor. Progress and preservation are not static; they are dynamic forces that require constant adjustment. Acting too slowly on climate change risks irreversible damage, while rushing into poorly planned development can create new problems. The balance lies in thoughtful action— responding to current needs while anticipating future consequences. This is where XY Logic plays a role: switching between modes of action and reflection, growth and stability, to maintain equilibrium.

Education is key to achieving this balance. People must understand that their choices, both individual and collective, impact the larger system. A farmer who adopts sustainable practices contributes to global food security. A company that reduces emissions sets a precedent for industry-wide change. A government that invests in renewable energy creates a ripple effect of innovation and adoption. Progress and preservation depend on informed decision-making at every level.

Ultimately, the balance of progress and preservation is about recognizing interconnectedness. Every action we take influences the system as a whole. When we act with awareness and responsibility, we ensure that progress does not come at the expense of what sustains us. This is not just a philosophical ideal but a practical necessity. The resources we depend on—clean air, water, and fertile land—are finite. Preserving them is not a choice; it is the foundation of all future progress.

The path forward requires collaboration. Governments, businesses, communities, and individuals must work together to align their goals. Policies that incentivize sustainability, technologies that prioritize efficiency, and behaviors that reduce waste are all parts of the solution. No single effort will suffice; the balance of progress and preservation requires a collective commitment.

In this chapter, we've explored the interplay between advancing and maintaining. The lesson is clear: progress without preservation is unsustainable, and preservation without progress is insufficient. Together, they form the dual forces that drive meaningful, lasting change. By embracing this balance, we can create a world where innovation and sustainability go hand in hand, ensuring a future that is not only brighter but also enduring.

Chapter 9
Timing as the Key to Interconnectedness

Timing governs everything. From the rhythm of the natural world to the intricate dynamics of human systems, timing determines outcomes. The sun rises and sets at predictable intervals, tides follow the pull of the moon, and even the smallest cells in our bodies operate on cycles. Yet, timing is often underestimated in how we plan, innovate, and act.

Understanding timing is essential because it is the invisible thread that weaves together the fabric of interconnectedness. Systems, whether natural or human-made, thrive or fail based on their ability to align with timing. Acting too soon can lead to inefficiency; acting too late can render even the best solutions irrelevant. The key is to find the right moment—a concept that applies to energy, economics, and even personal decisions.

Nature is the ultimate teacher of timing. Crops grow in specific seasons because they depend on the conditions of temperature, sunlight, and rainfall. Migratory birds travel thousands of miles, arriving precisely when food and weather conditions align with their survival. These systems work because they are attuned to time. When timing is disrupted—by climate change or human interference—entire ecosystems suffer.

In technology, timing is just as crucial. Consider renew- able energy. Solar panels capture energy only when the sun is shining. Conventional systems, such as inverters, optimize power output during peak sunlight hours but miss the energy available at dawn and dusk. The Switching Battery AC System, on the other hand, maximizes energy capture across the entire day, even in low-light conditions. By aligning with the natural rhythm of solar energy, it achieves higher efficiency and greater reliability. This is timing in action.

Timing also plays a critical role in economics. Investments made during economic downturns often yield the highest returns because they capitalize on recovery periods. Governments that act early to address climate change reduce long-term costs compared to those that delay. Timing decisions when to invest, innovate, or pivot can make or break businesses and nations. The most successful strategies are those that anticipate cycles and act at the right moment.

On a personal level, timing shapes our lives in profound ways. Relationships, careers, and even moments of inspiration depend on being ready and responsive to opportunities as they arise. The right action at the wrong time can fail; the wrong action at the right time can succeed. Mastering timing means developing an awareness of cycles, patterns, and the interconnected forces at play.

The concept of XY Logic ties directly to timing. By dynamically switching between parallel (X) and series (Y) configurations, the Switching Battery AC System adapts to varying conditions. This flexibility is a model for how we can approach timing in life. Instead of rigidly adhering to a single path, we must remain adaptable, shifting strategies as circumstances evolve. Timing is not about reacting but about anticipating knowing when to act and when to wait.

Global challenges demand a sense of timing. Climate change, for instance, requires urgent action. Delaying reduces options and increases costs, while acting too quickly without planning can lead to inefficiencies. The balance lies in thoughtful, immediate steps that align with long-term goals. Timing here is not just strategic; it is existential.

In technology, timing is just as crucial. Consider renewable energy. Solar panels capture energy only when the sun is shining. Conventional systems, such as inverters, optimize power output during peak sunlight hours but miss the energy available at dawn and dusk. The Switching Battery AC System, on the other hand, maximizes energy capture across the entire day, even in low-light conditions. By aligning with the natural rhythm of solar energy, it achieves higher efficiency and greater reliability. This is timing in action.

Timing also plays a critical role in economics. Investments made during economic downturns often yield the highest returns because they capitalize on recovery periods. Governments that act early to address climate change reduce long-term costs compared to those that delay. Timing decisions—when to invest, innovate, or pivot—can make or break businesses and nations. The most successful strategies are those that anticipate cycles and act at the right moment.

In energy systems, timing offers a blueprint for resilience. The ability to store energy when it is abundant and release it when it is scarce is critical to balancing supply and demand. The Switching Battery's ability to capture and deliver energy dynamically mirrors this principle. It shows that timing is not just about when energy is available but how systems adapt to make the most of it.

Education and awareness are key to mastering timing. Individuals and organizations must learn to recognize patterns and act in harmony with them. This means studying cycles—from market trends to natural seasons—and developing the flexibility to respond effectively. Timing is not about perfection but about alignment. Even small adjustments can yield significant results when they occur at the right moment.

Ultimately, timing is the rhythm of interconnectedness. It ensures that systems function cohesively, that resources are used efficiently, and that progress aligns with preservation. vWhen we understand timing, we see that it is not an external force but an integral part of how the world operates. By aligning our actions with the natural flow of time, we create harmony, reduce waste, and unlock the full potential of interconnected systems.

The lesson of timing is simple but profound: everything has its moment. By observing, understanding, and respecting timing, we can act with precision and purpose. This is not just a skill but a mindset—one that transforms how we live, work, and innovate. Timing is the key to unlocking the beauty and power of interconnectedness.

Chapter 10
Designing Systems for Oneness

In a world where everything is interconnected, the systems we design must reflect this reality. From energy grids to economies, education to governance, the way we build and operate our systems determines how well they align with the principle of oneness. A well-designed system supports interconnectedness, fosters mutual benefit, and promotes sustainability. A poorly designed one creates inefficiencies, conflicts, and destruction.

Designing systems for oneness starts with recognizing their purpose. Every system—whether technological, social, or economic—exists to serve the needs of the whole. This means that no system can function in isolation; it must work in harmony with others. The first step in this process is understanding the relationships between systems. How does one affect the other? How do they interact and depend on each other? Without this understanding, even the best-intentioned designs can fail.

Let's consider energy systems. Traditional grids were built with a one-way flow of power—from centralized power plants to consumers. This model ignored the potential of decentralized, renewable energy sources like solar and wind. As a result, it created inefficiencies and vulnerabilities.

The Switching Battery AC System represents a new approach, one that embraces interconnectedness. By integrating with renewable energy, eliminating the need for inverters, and enabling decentralized power generation, it demonstrates how systems can be designed to align with natural flows of energy. It doesn't just provide power; it creates resilience and adaptability within the larger grid.

This principle applies to other areas as well. In economics, systems that prioritize inclusivity and cooperation tend to be more stable and prosperous. Microfinance initiatives, for example, have shown how providing small loans to underserved communities can spark economic growth, reduce inequality, and build social trust. These systems succeed because they recognize that the economy is not a competition but a network where everyone's success contributes to the whole.

Education is another example. Traditional models often emphasize competition—ranking students, awarding limited opportunities, and rewarding individual achievement. But education systems designed for oneness prioritize collaboration and shared learning. They recognize that knowledge grows when it is shared and that the success of one student enriches the entire community. Such systems encourage creativity, empathy, and critical thinking,

56

preparing individuals not just for personal success but for collective progress.

Governance, too, must reflect interconnectedness. Policies that focus narrowly on short-term gains or national interests often lead to long-term problems and global instability. By contrast, systems of governance that consider the interconnected nature of climate, trade, migration, and public health are more effective at creating solutions that work for everyone. Multilateral agreements on issues like climate change and pandemic response demonstrate how cooperation produces better outcomes than isolation or conflict.

To design systems for oneness, we must also consider scalability. A system that works well at a small scale may fail when applied broadly unless it is designed with adaptability in mind. This is where concepts like XY Logic come into play. The Switching Battery AC System's ability to dynamically switch between configurations is a metaphor for how systems must operate—adapting to changing conditions, scaling up or down as needed, and remaining flexible in the face of uncertainty.

Another critical aspect of system design is feedback. No system is perfect from the outset. To align with oneness, systems must be designed to learn and evolve. This requires mechanisms for monitoring, feedback, and adjustment. In natural ecosystems, feedback loops maintain balance—predator and prey populations, for instance, adjust to each other over time. Human systems must do the same, using data, analysis, and input from stakeholders to continuously improve.

Designing for oneness also involves embracing simplicity. Complexity can lead to inefficiency and confusion, while simplicity allows systems to function more effectively and inclusively. The Switching Battery's design exemplifies this principle. By eliminating unnecessary components like inverters and relying on straightforward, scalable configurations, it achieves greater efficiency and accessibility. Simplicity does not mean lack of sophistication; it means clarity of purpose and elegance in execution.

Finally, designing systems for oneness requires a shift in values. It means prioritizing collaboration over competition, sustainability over short-term gain, and equity over exclusion. These values must be embedded in the very fabric of the systems we create. They must guide decisions at every level, from policy-making to technological innovation.

The systems we design shape the world we live in. When we design them with interconnectedness in mind, we create a foundation for harmony, resilience, and progress. By recognizing that every system is part of a larger whole, we move closer to a world where oneness is not just a principle but a reality. This is the ultimate goal of designing systems for oneness: to create a world where everyone thrives together.

Chapter 11
The Human Element in Interconnectedness

At the heart of every system, invention, and idea lies the human element. Machines can calculate, automate, and optimize, but it is human values, creativity, and purpose that drive progress. The interconnectedness of life depends not just on physical systems but on how people interact, collaborate, and contribute to the whole. Understanding and embracing this human element is crucial for building a future that works for everyone.

Humans are unique in their ability to imagine and innovate. This capacity has allowed us to create technologies that amplify our abilities, from the wheel to the internet. But with this power comes responsibility. The choices we make about how to use our tools and resources have profound consequences. Will we design systems that promote equity and sustainability, or will we continue to prioritize short-term gains at the expense of others?

One of the most critical aspects of the human element is empathy. Empathy allows us to see the world from another's perspective, to understand their needs, and to act in ways that benefit not just our selves but others. In an interconnected world, empathy is not optional; it is essential. Without it, systems become rigid, self-serving, and prone to failure. With it, we create networks of mutual support that strengthen the whole.

Collaboration is another defining feature of humanity. While competition has driven much of our history, collaboration is what enables true progress. Great achievements—whether building the pyramids, developing vaccines, or exploring space—are the result of people working together toward a common goal. In today's world, the challenges we face—climate change, inequality, and technological disruption—are too complex for any one person, organization, or nation to solve alone. Collaboration is not just a choice; it is a necessity.

The Switching Battery AC System is an example of how the human element transforms technology. The system is not just a technical achievement; it is a solution designed with people in mind. Its affordability, adaptability, and simplicity reflect a deep understanding of the needs of underserved communities. It bridges the gap between cutting-edge innovation and practical application, showing how technology can empower rather than exclude.

The human element also includes creativity. Creativity allows us to see connections that others miss, to imagine possibilities beyond the status quo, and to find solutions where none seem to exist. It is the spark that drives innovation and the glue that binds interconnected systems. Creativity thrives in environments where people feel safe, valued, and inspired. By fostering such environments, we unlock the full potential of human ingenuity. However, the human element is also the source of many challenges. Bias, fear, and short-sightedness can undermine interconnectedness. When people act out of self-interest or mistrust, they weaken the systems that support us all. Addressing these challenges requires intentional effort to cultivate trust, transparency, and shared purpose. Education, dialogue, and leadership play vital roles in shaping a culture that values interconnectedness.

Timing, a theme explored earlier, also applies to the human element. Knowing when to speak, when to listen, and when to act is as important in human interactions as it is in technological systems. Effective leaders, educators, and innovators understand this. They align their actions with the needs of the moment, creating momentum for positive change.

Ultimately, the human element is what makes interconnectedness meaningful. Systems and technologies are tools; it is people who give them purpose. Our ability to imagine, create, and care for one another is what transforms a collection of parts into a living, thriving whole. By embracing the human element, we not only improve our systems but also enrich our lives.

The future depends on how we harness the human element in interconnectedness. Will we use our empathy, creativity, and collaboration to build systems that uplift everyone? Or will we allow fear, greed, and division to fracture the whole? The choice is ours, and the stakes could not be higher. By recognizing the human element as central to everything we do, we can create a world that is not just interconnected but truly unified.

Chapter 12
Toward a Unified Future

The future is not something we enter; it is something we create. Each decision we make today shapes the world of tomorrow. As we move forward, the vision of a unified future—one that embraces interconnectedness, sustainability, and oneness—must guide our actions. This chapter explores how we can build that future, drawing on the principles and ideas discussed throughout this book.

1. **Embracing Interconnectedness in Decision-Making**

The first step toward a unified future is recognizing that every decision we make impacts the larger system. Whether it's an individual choosing to reduce energy consumption or a government setting policies on renewable energy, each choice sends ripples through the interconnected web of life. By considering the long-term consequences and interdependencies of our actions, we can make decisions that benefit not just ourselves but the whole.

For example, cities that invest in renewable energy infrastructure not only reduce their carbon footprint but also create jobs, improve public health, and enhance resilience against climate disruptions. Similarly, businesses that adopt sustainable practices often find that they attract loyal customers and reduce costs in the long run. These outcomes demonstrate the power of interconnected thinking.

2. Building Inclusive Systems

A unified future cannot exist without inclusivity. Systems that exclude or marginalize certain groups are inherently unstable and unjust. To create systems that truly reflect oneness, we must design for diversity and equity. This means ensuring access to resources like education, healthcare, and clean energy for everyone, regardless of geography or socioeconomic status.

The Switching Battery AC System exemplifies this principle. By making clean energy affordable and accessible, it empowers individuals and communities that have traditionally been left out of the energy transition. This model can be replicated in other areas, from water distribution to education, creating systems that lift everyone rather than a select few.

3. Prioritizing Sustainability

Sustainability is the cornerstone of a unified future. Without it, progress is short-lived and ultimately destructive. Sustainability requires balancing the needs of the present with the needs of future generations. This involves not only protecting natural resources but also developing technologies and systems that operate in harmony with the planet's cycles.

Renewable energy is a critical component of this vision. By transitioning away from fossil fuels and toward decentralized, renewable systems, we can reduce emissions, minimize waste, and create a more stable energy supply. The principles of XY Logic, which allow for adaptability and efficiency, provide a framework for designing such systems.

4. Fostering Collaboration Across Boundaries

A unified future demands collaboration at every level. Nations must work together to address global challenges like climate change, pandemics, and inequality. Businesses and governments must collaborate to create policies and innovations that benefit both the economy and society. Individuals must work together to build stronger communities and networks of support.

Collaboration is not just a practical necessity; it is a reflection of our interconnected nature. When we work together, we amplify our collective strengths and mitigate individual weaknesses. The most effective solutions arise when diverse perspectives and talents come together toward a shared goal.

5. Cultivating a Culture of Oneness

Ultimately, a unified future requires a cultural shift. We must move beyond the mindset of competition and scarcity and embrace one of abundance and cooperation. This involves rethinking how we define success, shifting from individual achievement to collective well-being.

Education plays a vital role in this transformation. By teaching interconnectedness and systems thinking from an early age, we can equip future generations with the tools they need to navigate complexity and build a better world. Media, art, and storytelling can also inspire this shift by highlighting stories of collaboration, resilience, and shared humanity.

The Path Forward

A unified future is not a distant dream; it is an achievable reality. By embracing interconnectedness, designing inclusive and sustainable systems, fostering collaboration, and cultivating a culture of oneness, we can create a world where everyone thrives. The journey will not be easy, but it is necessary. Each step we take brings us closer to a future that reflects the true nature of life: interconnected, interdependent, and indivisible.

The time to act is now. The choices we make today will determine the legacy we leave for generations to come. Let us choose unity, sustainability, and hope. Let us build a future that honors the interconnectedness of all things and creates a world where progress and preservation go hand in hand.

Chapter 13
The Power of Hope and Action

Hope is not a passive emotion; it is a catalyst for change. When combined with action, hope has the power to transform lives, systems, and societies. In a world facing challenges like climate change, inequality, and resource depletion, hope is the driving force that enables us to envision a better future and take the necessary steps to achieve it.

Hope begins with belief—the belief that change is possible. Throughout history, transformative progress has been driven by individuals and communities who dared to imagine what others thought impossible. From civil rights movements to technological breakthroughs, every major advancement started with a vision of a better world. Hope provides the energy to pursue that vision, even when the odds seem insurmountable.

However, hope alone is not enough. It must be paired with action. Action gives hope its power by turning ideas into reality. Without action, hope becomes wishful thinking. The key is to align hope with practical steps that lead to tangible outcomes. This alignment requires planning, commitment, and the courage to act, even in the face of uncertainty.

The Switching Battery AC System embodies this principle. It started as a vision: a more efficient, affordable, and sustainable way to generate and distribute energy. But it didn't remain an idea. Through research, development, and collaboration, it became a functional system that is changing how energy is produced and consumed. The system represents what is possible when hope meets action, and it serves as a model for tackling other challenges.

Hope is also contagious. When people see progress, it inspires them to contribute and collaborate. A single innovation can spark a ripple effect, encouraging others to innovate in their own ways. For example, the adoption of renewable energy technologies has inspired a global movement toward sustainability, with individuals, businesses, and governments all playing a role. Hope grows when it is shared, creating a collective momentum that drives meaningful change.

Hope must however be grounded in reality. False hope—the belief that solutions will emerge without effort or sacrifice—can lead to complacency. Real hope acknowledges the challenges ahead while remaining focused on the possibilities. It requires honesty about what needs to change and a willingness to confront difficult truths. This balance between optimism and realism is what makes hope a powerful and effective force. Action, like hope, thrives on collaboration. No individual or organization can solve global challenges alone.

Progress requires partnerships that bring together diverse skills, perspectives, and resources. Whether it's governments working on international agreements, businesses developing innovative solutions, or communities coming together to address local issues, collective action amplifies the impact of hope.

Timing is also critical. Acting too soon without preparation can lead to failure, while waiting too long can result in missed opportunities. The interplay between hope and action requires an understanding of when to act and how to build momentum. This is where concepts like XY Logic come into play, providing the flexibility to adapt and respond to changing circumstances while staying aligned with long-term goals.

Hope and action also require persistence. Change is rarely immediate, and progress often comes in small steps rather than giant leaps. Setbacks are inevitable, but they should not deter us. Every challenge is an opportunity to learn, adapt, and improve. Persistence ensures that hope remains alive and that action continues, even in the face of adversity.

As we look toward the future, hope and action must become guiding principles in everything we do. They must drive our decisions, shape our systems, and inspire our communities. Whether it's addressing climate change, reducing inequality, or creating new technologies, the combination of hope and action is what will enable us to build a better world.

The lesson is clear: hope provides the vision, and action makes it real. Together, they are the forces that drive progress and transformation. By embracing these principles, we can overcome the challenges we face and create a future that reflects the interconnectedness, sustainability, and unity we aspire to achieve.

Chapter 14
Beyond the False Dichotomy

Human society often simplifies complexity into opposite: "good or bad", "winner or loser", or "love or hate". While this binary thinking provides clarity in some situations, it also traps us in a cycle of division. By framing decisions as zero-sum games—where one person's gain is another's loss—we miss opportunities for collaboration, innovation, and shared success.

Beyond the False Dichotomy is a philosophy of balance, harmony, and openness to possibilities. It challenges us to reject the rigidity of extremes and explore the continuum of choices that lie between them. This is not just about altruism; it is a practical and rewarding approach that creates mutual benefit. By recognizing that our own well-being is tied to the well-being of others, we unlock the potential for a better, more prosperous world.

The Neighborhood View: A Metaphor for Shared Prosperity

Imagine a neighborhood where I live in a large, luxurious home, while my neighbor resides in a modest, run-down house. From my window, I see his dilapidated property every day—a view that detracts from the beauty of my surroundings. Meanwhile, my neighbor enjoys a "rich view," looking out at my grand home.

This dynamic illustrates an important truth: the quality of my environment, even as a wealthy individual, is deeply influenced by the condition of my neighbor's property. If I want a better view, I must help my neighbor improve his home. By investing in my neighbor's well-being, I enhance the value of my own property. This is not just moral—it is practical.

When my neighbor prospers, the entire neighborhood thrives. Property values rise, crime rates fall, and the community becomes more vibrant. This principle applies not just to neighborhoods but to countries, regions, and the world at large. Every nation that has achieved prosperity has done so by fostering collective progress, ensuring that no one is left behind. Shared success strengthens the whole system.

This is the essence of Beyond the False Dichotomy: helping others is not an act of charity—it is an investment in mutual prosperity.

Binary Thinking and the Zero-Sum Trap

Binary thinking—framing the world in opposites like success versus failure or right versus wrong—has shaped human society for centuries.

While this mindset simplifies complexity, it often hides the nuanced interconnections between opposing forces.

At the heart of this binary thinking lies the zero-sum game, which assumes that resources, opportunities, and success are finite. This mindset fosters competition and conflict, limiting progress in many areas:

1. **Energy Systems:** Fossil fuels and renewables are often framed as opposites, forcing a choice between the two. This false dichotomy ignores the transitional period needed to integrate renewables while maintaining reliable energy.

2. **Business:** Companies feel pressured to choose between profit and purpose, missing the opportunity to align the two for long-term success.

3. **Social Structures:** Individual success is pitted against collective welfare, creating divisions that weaken communities and societies.

The result is a world trapped in conflict, where the potential for win-win solutions is overlooked.

Beyond the Zero-Sum Game: A Positive-Sum Approach

The philosophy of Beyond the False Dichotomy rejects the zero-sum mindset in favor of a positive-sum game. It recognizes that progress is not a finite resource and that collaboration creates outcomes far greater than the sum of their parts.

For instance, XY Logic, a revolutionary framework in energy systems, replaces the inefficiencies of binary on-off switching with dynamic modulation between parallel (X) low voltage and series (Y) high voltage configurations. This approach eliminates wasteful energy spikes, increasing efficiency and creating win-win solutions.

Applied to society, XY Logic illustrates how those with more resources (Y) can uplift those with fewer resources (X) without diminishing their own value. By sharing a fraction of their wealth or opportunities, the wealthy not only empower the underserved but also create stronger, more resilient systems that benefit everyone.

Balance and Harmony: The Key to Unlocking Possibilities

Beyond the False Dichotomy is fundamentally about balance and harmony. It challenges us to move beyond extremes and explore the continuum of choices that exist between them. Why settle for a choice between two polarities when there are a hundred possibilities that might better serve us?

This requires openness to possibilities (OTP). We must:

- Recognize that we are entitled to explore all options.

74

- Examine the full spectrum of choices to identify those that are actionable and aligned with our goals.

- Shift from a mindset of competition to one of collaboration, creating solutions that benefit everyone.

Balance is not about compromise—it is about integration. It is about finding synergy where opposing forces complement one another, creating outcomes that are greater than their individual contributions.

Lessons from Nature

Nature offers profound examples of balance, harmony, and interconnectedness:

1. **The Water Cycle:** Water transitions between vapor, liquid, and ice, sustaining ecosystems at every stage. Each phase depends on the others, forming a continuous loop of renewal.

2. **Photosynthesis and Respiration:** Plants and humans exchange gases in perfect balance—carbon dioxide and oxygen—demonstrating how collaboration sustains life.

3. **Ecosystem Dynamics:** Predators and prey maintain population balance, supporting biodiversity.

These natural systems remind us that opposites are not adversaries but complementary forces. They teach us that harmony, not competition, is the foundation of progress.

The Amber Mode: Bridging Divides

The Amber Mode, a feature of the Switching Battery, provides a powerful metaphor for moving beyond extremes. Just as the amber light in a traffic signal bridges the stop (red) and go (green) states, the Amber Mode allows for smooth transitions between low and high voltages in energy systems.

This transitional state embodies the philosophy of Beyond the False Dichotomy: it creates a space for balance and harmony, enabling dynamic movement between opposites without disruption or conflict.

Practical Applications

Energy Systems: Technologies like the Switching Battery show that renewable energy can coexist with existing infrastructure during the transition to sustainability.

- **Business Models:** Companies like Patagonia prove that aligning profit with purpose creates long-term value for shareholders and society.

- **Social Equity:** Uplifting underserved communities benefits everyone. A healthier, more educated population drives economic growth and innovation.

- **Personal Growth:** Balancing ambition with gratitude, rest with productivity, and individuality with community leads to a more fulfilling life.

Redefining Success

True success is not a zero-sum game. It is a journey of positive-sum thinking, where individual and collective progress reinforce one another. This perspective:

- Encourages us to redefine wealth as more than material accumulation.

- Highlights the rewards of sharing, which enrich the giver as much as the receiver.

- Demonstrates that collaboration, not competition, drives lasting progress.

When we uplift others, we uplift ourselves. Helping our neighbors improve their lives increases the value of our own. This principle applies at every scale—from neighborhoods to nations to the world.

The Power of Balance and Harmony

Beyond the False Dichotomy challenges us to reject artificial divisions and embrace the interconnectedness of life. It invites us to see the world not as a battle between extremes but as a continuum of possibilities where balance and harmony create shared prosperity.

The lesson is clear: when we invest in others, we invest in ourselves. By finding the middle ground, exploring all possibilities, and embracing the power of oneness, we unlock a future where everyone thrives.

Chapter 15
The Path to Collective Abundance

Abundance is often misunderstood. Many view it as the accumulation of wealth, resources, or power. But true abundance is not about excess; it is about balance, access, and shared prosperity. In an interconnected world, abundance is not something to hoard but something to create and distribute collectively. This chapter explores how we can shift from a mindset of scarcity to one of abundance, unlocking the full potential of humanity and our planet.

The Illusion of Scarcity

Scarcity is often used to justify competition and inequality. We are told that there isn't enough to go around, that we must fight for limited resources, and that some must lose for others to win. But this narrative is rarely accurate. In many cases, scarcity is artificially created by inefficient systems, inequitable distribution, or short-term thinking.

Consider food. The world produces enough to feed everyone, yet hunger persists because of waste, poor logistics, and economic barriers. Similarly, the sun provides more energy in a single hour than humanity uses in a year, yet billions lack access to reliable electricity. The problem is not a lack of resources; it is a failure to manage and distribute them effectively.

Redefining Abundance

True abundance is about sufficiency for all, not excess for a few. It is about designing systems that meet everyone's needs while respecting the planet's limits. This requires a fundamental shift in how we think about wealth, progress, and success. Abundance is not a static state; it is a dynamic balance where resources flow efficiently and equitably.

This principle is reflected in the Switching Battery AC System. By harnessing renewable energy and eliminating wasteful processes, it creates more with less. It doesn't just provide power; it democratizes energy access, enabling communities to thrive without overburdening the environment. This is the essence of collective abundance: creating systems that uplift everyone while preserving the resources that sustain us.

The Role of Collaboration

Abundance is inherently collective. It cannot be achieved in isolation. Collaboration is the key to unlocking shared prosperity, whether through partnerships between nations, businesses, or communities. When we pool resources, knowledge, and effort, we multiply our capacity to create value.

80

For example, global initiatives to share renewable energy technologies can accelerate the transition to a sustainable future. Local cooperatives that manage resources collectively, such as community solar projects, demonstrate how collaboration can generate economic and social benefits. These models show that abundance grows when it is shared.

Overcoming Barriers to Abundance

Achieving collective abundance requires addressing the barriers that perpetuate scarcity. These include:

- **Inequitable Systems:** Wealth and power are often concentrated in ways that exclude others. Reforms in taxation, governance, and corporate practices can redistribute resources more fairly.

- **Inefficiencies:** Many systems waste resources due to outdated technologies or misaligned incentives. Investing in innovation and redesigning processes can unlock untapped potential.

- **Mindset:** Perhaps the greatest barrier is the belief that scarcity is inevitable. Changing this mindset requires education, storytelling, and leadership that inspire hope and possibility.

A Vision for the Future

Imagine a world where abundance is the norm. Clean energy powers homes and industries without depleting natural resources. Food systems eliminate waste and ensure that everyone is nourished. Communities thrive because they share resources, skills, and opportunities. This vision is not utopian; it is achievable if we align our systems and values with the principles of interconnectedness and equity.

The journey to collective abundance starts with small steps. Each decision to share rather than hoard, to innovate rather than settle, and to collaborate rather than compete brings us closer to this vision. Whether it's a government investing in renewable energy, a company prioritizing sustainability, or an individual supporting local initiatives, every action matters.

The Abundance Mindset

Abundance is not just about resources; it is about relationships. It is about how we treat each other and our planet. When we move from a mindset of scarcity to one of abundance, we see that there is enough for everyone—if we work together to create and distribute it.

The path to collective abundance is both practical and profound. It challenges us to rethink our systems, our values, and our actions. But it also offers hope: the promise of a world where everyone has the opportunity to thrive. By embracing this vision, we can build a future that reflects the interconnectedness and unity of life, ensuring that abundance is not just possible but inevitable for all.

Chapter 16
Leadership for an Interconnected World

Leadership has always been a defining force in shaping societies, systems, and the future. But in an interconnected world, leadership must evolve. It is no longer about command and control or individual power; it is about guiding collective action, fostering collaboration, and creating systems that reflect the oneness of life. This chapter explores the qualities, principles, and practices of effective leadership in an interconnected age.

Redefining Leadership

Traditional leadership often focuses on hierarchy and authority. Leaders are seen as decision-makers who guide others through directives. While this approach has its place, it falls short in addressing the complexities of today's challenges. In a world where everything is connected, effective leadership requires inclusivity, adaptability, and vision.

The new model of leadership is about empowerment. Instead of dictating actions, leaders facilitate collaboration, ensuring that diverse voices are heard and valued. They focus on building trust, aligning goals, and creating environments where people can contribute their best work. This type of leadership recognizes that no one person has all the answers; solutions emerge from collective effort.

The Qualities of an Interconnected Leader

- **Empathy:** Understanding the needs, perspectives, and emotions of others is crucial for fostering collaboration and building trust. Empathetic leaders listen actively and create spaces where people feel respected and valued.

- **Vision:** Leaders must inspire others with a clear and compelling vision of the future. This vision should reflect interconnectedness, showing how individual efforts contribute to collective goals.

- **Adaptability:** In a rapidly changing world, rigidity is a liability. Leaders must be flexible, willing to adjust strategies as circumstances evolve. This requires openness to feedback and a willingness to learn.

- **Collaboration:** The best leaders know how to bring people together. They facilitate teamwork, mediate conflicts, and ensure that everyone's contributions are integrated into the larger effort.

- **Integrity:** Trust is the foundation of effective leadership. Leaders must act with honesty, fairness, and consistency, aligning their actions with their values and commitments.

Leading by Example

Leadership is not about telling others what to do; it is about showing them what is possible. Leaders must embody the principles they advocate. If sustainability is a priority, they should adopt sustainable practices. If inclusivity is a goal, they should ensure that their own actions reflect it. Leading by example inspires others to follow and creates a culture of accountability.

Systems Thinking in Leadership

An interconnected world demands systems thinking. Leaders must understand how different elements of a system interact and influence one another. This perspective helps identify root causes of problems and design solutions that address them holistically. For example, a leader addressing climate change must consider not only emissions reduction but also economic impacts, social equity, and technological innovation.

The Switching Battery AC System is a case in point. Its development required systems thinking to integrate renewable energy, battery storage, and grid connectivity. The result is a solution that aligns technological innovation with environmental sustainability and social benefit. Leaders in other fields can learn from this approach, designing systems that work seamlessly across boundaries.

Leadership in Action

Real-world leadership involves making difficult decisions, managing uncertainty, and navigating competing interests. The following practices can help leaders succeed:

- **Facilitating Collaboration:** Bring diverse stakeholders together to co-create solutions. This includes listening to underrepresented voices and ensuring that decisions reflect a broad range of perspectives.

- **Setting Clear Goals:** Provide a clear direction while allowing flexibility in how goals are achieved. This balance empowers teams while maintaining focus.

- **Communicating Effectively:** Transparency and clarity are essential. Leaders must articulate their vision, explain their decisions, and address concerns openly.

- **Encouraging Innovation:** Create environments where experimentation is encouraged and failure is seen as a learning opportunity. This fosters creativity and resilience.

A New Paradigm of Leadership

Leadership in an interconnected world is not confined to titles or positions. It is a mindset and a practice that anyone can adopt. Whether you are leading a team, a community, or simply your own life, the principles of interconnected leadership apply. By acting with empathy, vision, and integrity, you contribute to the larger system and inspire others to do the same.

This new paradigm recognizes that leadership is not about control but about connection. It is about guiding others toward shared goals, building systems that reflect oneness, and creating a future where everyone thrives. In an interconnected world, leadership is not just a role; it is a responsibility—one that we all share.

Chapter 17
The Responsibility of Knowledge

Knowledge is both a gift and a responsibility. In an interconnected world, the information we possess shapes not only our individual lives but also the systems, communities, and environments we influence. How we use knowledge—whether to create, empower, or exploit— determines the trajectory of progress and the sustainability of our shared future. This chapter explores the ethical and practical responsibilities that come with knowledge and how embracing them can lead to transformative change.

The Power of Knowledge

Knowledge is the foundation of human progress. It has allowed us to harness energy, build civilizations, and explore the universe. But the power of knowledge is not neutral. Like energy, it can be used constructively or destructively. A technology that improves lives can also create inequalities if access is limited. An innovation that addresses one problem can create unintended consequences elsewhere. Recognizing this duality is the first step toward using knowledge responsibly.

89

In the development of the Switching Battery AC System, knowledge played a central role. The system was born from an understanding of energy flows, battery dynamics and renewable integration. But its true power lies in how this knowledge was applied: not to maximize profits for a few but to create an affordable, sustainable solution for many. This is an example of knowledge used responsibly, aligned with the principles of interconnectedness and equity.

Ethical Stewardship of Knowledge

With knowledge comes ethical responsibility. Those who hold knowledge—whether scientists, policymakers, or educators—must consider the broader impact of their actions. Ethical stewardship involves asking questions such as:

- **Who benefits?** Does this knowledge serve the many or the few?

- **What are the risks?** Are there unintended consequences, and how can they be mitigated?

- **Is it inclusive?** Does this knowledge empower marginalized communities, or does it widen gaps?

Ethical stewardship requires a balance between innovation and caution. It means pursuing breakthroughs while ensuring they align with shared values and long-term sustainability.

The Democratization of Knowledge

For knowledge to create collective progress, it must be accessible. Hoarding information—whether through patents, paywalls, or exclusive systems—limits its potential. The democratization of knowledge ensures that ideas and innovations benefit everyone, not just a privileged few.

Open-source platforms, collaborative research, and equitable education systems are examples of democratized knowledge in action. By sharing what we know, we accelerate progress, foster innovation, and build trust. The Switching Battery's affordability and simplicity reflect this principle, making advanced energy solutions accessible to underserved communities.

Knowledge as a Catalyst for Empowerment

Knowledge is most powerful when it empowers others. Education, for example, is not just about transferring information but about giving people the tools to shape their own lives and communities. Empowerment through knowledge builds resilience, reduces dependency, and unlocks creativity.

In energy systems, this principle is evident in decentralized solutions like the Switching Battery. By enabling individuals and communities to generate and manage their own energy, these systems reduce reliance on centralized grids and empower users to take control of their energy future. Knowledge, in this context, becomes a tool for liberation.

The Risks of Ignoring Responsibility

Failing to use knowledge responsibly has direct consequences. Technologies developed without considering their broader impact can harm ecosystems, destabilize economies, and deepen inequalities. Historical examples abound, from industrial pollution to the unintended consequences of artificial intelligence.

In an interconnected world, the ripple effects of irresponsibility are magnified. A single decision can trigger global consequences, underscoring the need for thoughtful, informed action. The more powerful the knowledge, the greater the responsibility to use it wisely.

A Vision for Responsible Knowledge

A future where knowledge is used responsibly is a future where progress aligns with sustainability, equity, and interconnectedness. Achieving this vision requires:

- **Ethical Guidelines:** Establishing clear principles to guide the development and application of knowledge.

- **Collaboration:** Sharing knowledge across disciplines, sectors, and borders to create holistic solutions.

- **Education:** Equipping people with the skills to critically evaluate and apply knowledge responsibly.

- **Transparency:** Ensuring that decisions involving knowledge are open and inclusive.

Knowledge as a Shared Asset

Knowledge is not a commodity to be owned; it is a shared asset that connects and uplifts humanity. By embracing the responsibility that comes with knowledge, we can harness its power to address global challenges, foster innovation, and create systems that reflect the oneness of life.

The path forward requires humility, foresight, and a commitment to the greater good. Knowledge, when used responsibly, is the bridge between what is and what could be. It is the key to a future where interconnectedness is not just understood but lived, ensuring that progress benefits everyone and sustains the world we all share.

Chapter 18
The Beauty of Simplicity

In a world of increasing complexity, simplicity stands out as a powerful force. Simplicity is not about reducing depth or eliminating nuance; it is about clarity, purpose, and efficiency. It is about stripping away the unnecessary to reveal what truly matters. In an interconnected world, simplicity becomes a guiding principle that fosters understanding, innovation, and harmony.

The Misconception About Simplicity

Simplicity is often misunderstood as a lack of sophistication or depth. In reality, achieving simplicity requires effort, insight, and mastery. The simplest solutions are often the most elegant because they distill complexity into something clear and actionable. Einstein's famous quote, "Everything should be made as simple as possible, but not simpler," captures this essence.

Nature exemplifies simplicity. A tree, for instance, performs a complex series of functions—photosynthesis, water transport, and oxygen production—through straightforward structures like leaves,

The Power of Oneness

roots, and branches. These processes work seamlessly because they are efficient and purpose-driven. Similarly, human systems must aim for simplicity without losing the intricacies that make them effective.

Simplicity in Technology

The Switching Battery AC System demonstrates the beauty of simplicity in technology. By eliminating inverters and transformers, it reduces inefficiencies and complexity. Its design focuses on what is essential: harnessing energy effectively and delivering it reliably. This simplicity makes it not only efficient but also accessible, ensuring that more people can benefit from clean energy solutions.

This principle applies to all technologies. The most successful innovations are those that simplify the user experience while addressing real needs. Smartphones, for example, consolidate multiple tools into a single device, streamlining communication, navigation, and information access. The challenge for designers and engineers is to simplify without oversimplifying—to preserve functionality while enhancing usability.

Simplicity in Systems

Complex systems often fail because their intricacies create inefficiencies and vulnerabilities. Simplicity in systems, whether in governance, education, or healthcare, allows for greater adaptability

96

and resilience. A well-designed system focuses on core objectives and aligns its components to achieve them efficiently.

In education, for instance, simplicity can mean focusing on foundational skills like critical thinking, communication, and problem-solving rather than overloading students with information. In governance, it can mean creating policies that are transparent and easy to implement, reducing bureaucracy and fostering trust. Simplicity does not mean ignoring complexity; it means managing it effectively.

The Role of Simplicity in Communication

Interconnectedness requires clear communication, and simplicity is its cornerstone. When ideas are expressed simply, they become more accessible and actionable. Complex jargon and convoluted explanations create barriers to understanding and collaboration.

The ability to communicate complex ideas in simple terms is a hallmark of great leaders and thinkers. It bridges gaps between disciplines, cultures, and perspectives, enabling shared understanding and collective action. Whether explaining a new technology or advocating for social change, simplicity in communication is essential for building connections and driving progress.

The Art of Simplification

Simplification is an art that requires:

- **Clarity of Purpose:** Understanding what truly matters and focusing efforts accordingly.

- **Elimination of Redundancy:** Removing elements that do not contribute to the core goal.

- **Iterative Refinement:** Continuously improving designs, systems, or messages to enhance their simplicity and effectiveness.

- **Empathy:** Considering the needs and perspectives of those who will interact with the system or idea.

Simplicity as a Value

Simplicity is not just a design principle; it is a value that can transform how we live and work. Embracing simplicity means prioritizing quality over quantity, clarity over noise, and purpose over distraction. It means designing lives, systems, and technologies that are not just functional but meaningful.

In a world overwhelmed by information, options, and complexity, simplicity offers a path to focus and balance. It helps us navigate interconnected systems with confidence and clarity, ensuring that our actions align with our goals and values.

The Power of Simplicity

Simplicity is the ultimate sophistication. It reveals the essence of things, making the complex understandable and the overwhelming manageable. In an interconnected world, simplicity is not a luxury but a necessity. It enables us to build systems that work, communicate ideas that resonate, and create technologies that empower.

By embracing simplicity, we honor the interconnectedness of life. We focus on what truly matters, creating space for innovation, collaboration, and harmony. In this simplicity, we find not only beauty but also the key to progress and sustainability.

Chapter 19
The Circle of Reciprocity

In an interconnected world, every action creates ripples. What we give to the world—be it resources, energy, or kindness—inevitably returns to us in some form. This is the essence of reciprocity, a principle that governs ecosystems, economies, and human relationships. Understanding and embracing this circle of reciprocity is key to creating systems and societies that are balanced, sustainable, and just.

Reciprocity in Nature

Nature operates on cycles of reciprocity. Plants provide oxygen and food, while animals return nutrients to the soil. Rivers nourish the land, and rain replenishes the rivers. These exchanges sustain life, creating a balance that allows ecosystems to thrive. When this balance is disrupted— through deforestation, pollution, or overextraction—the entire system suffers.

The lesson is clear: reciprocity is not just an ideal but a necessity. By giving back as much as we take, we ensure the longevity and health of the systems that sustain us. This principle can guide how we approach not only environmental challenges but also social and economic systems.

Reciprocity in Human Systems

Human systems—whether economic, social, or technological—function best when they are rooted in reciprocity. A healthy economy, for example, relies on a balance between production and consumption. Businesses thrive when they reinvest in their workers, customers, and communities. Governments succeed when they prioritize public welfare, creating trust and stability in return.

Social systems also depend on reciprocity. Communities flourish when people support one another, sharing resources, time, and care. Acts of kindness and generosity create a ripple effect, inspiring others to do the same. This reciprocity builds networks of trust and mutual support, making communities more resilient and inclusive.

The Role of Technology in Reciprocity

Technology can amplify or disrupt reciprocity, depending on how it is designed and used. The Switching Battery AC System is an example of technology that aligns with the principle of reciprocity. By enabling decentralized energy generation and reducing waste, it gives back to the environment while empowering individuals and communities. It demonstrates that innovation can serve as a bridge, creating systems that are both efficient and regenerative.

However, not all technologies follow this principle. Systems that extract resources without replenishing them or prioritize profit over sustainability break the circle of reciprocity. These systems may offer short-term benefits but create long-term imbalances, harming both people and the planet.

Reciprocity as a Guiding Value

Embracing reciprocity as a guiding value requires a shift in mindset. It means viewing every interaction—whether with nature, technology, or other people—as part of a larger cycle. This perspective encourages us to act with intention, ensuring that what we give aligns with what we hope to receive.

In practice, this can take many forms:

- **Sustainable Practices:** Using resources responsibly and replenishing what we take, whether through renewable energy, reforestation, or circular economies.

- **Fair Exchange:** Ensuring that transactions—from trade agreements to workplace policies—are equitable and beneficial to all parties involved.

- **Community Engagement:** Contributing time, skills, or resources to strengthen the social fabric, knowing that a strong community benefits everyone.

The Ripple Effect of Reciprocity

One of the most powerful aspects of reciprocity is its ripple effect. A single act of giving can inspire others to do the same, creating a chain reaction of positive impact. This is true on both small and large scales. A neighbor helping a neighbor can transform a community, just as a nation sharing technology or resources can uplift entire regions.

The Switching Battery's potential to transform energy access exemplifies this ripple effect. By providing affordable, sustainable energy solutions, it empowers individuals and communities, creating opportunities for education, entrepreneurship, and improved quality of life. These benefits extend outward, contributing to economic growth and environmental preservation.

Living the Circle

Reciprocity is not just a principle; it is a way of life. It reminds us that we are part of a larger whole and that our actions have consequences beyond ourselves. By embracing the circle of reciprocity, we create systems and relationships that are balanced, sustainable, and just.

In an interconnected world, reciprocity is the foundation of progress and harmony. It ensures that what we build today supports the needs of tomorrow, creating a future where everyone thrives. The circle of reciprocity is not just a concept to understand but a value to live by, guiding our actions and shaping a world that reflects the beauty of interconnectedness.

Chapter 20
The Future We Create Together

The future is not predetermined; it is shaped by the choices we make today. In an interconnected world, where every action sends ripples through systems and societies, our collective decisions hold immense power. By reflecting on how we can harness the principles of interconnectedness, reciprocity, and sustainability to build a future that reflects the best of humanity.

The Power of Collective Action

Individual actions matter, but their impact is magnified when aligned with collective purpose. Movements, innovations, and revolutions are born when people come together to address shared challenges. The energy transition is a prime example. While one person installing solar panels or using sustainable energy may seem small, the collective adoption of renewable technologies has the potential to transform global energy systems.

The Switching Battery AC System exemplifies how individual and collective actions intersect. It empowers individuals to generate and store their own energy, reducing dependence on centralized grids. When adopted widely, it contributes to a decentralized, sustainable energy future. This is the power of collective action: small changes adding up to systemic transformation.

Designing a Unified Vision

A unified vision for the future requires collaboration across sectors, disciplines, and borders. Governments, businesses, and communities must work together to align goals and resources. This vision should prioritize equity, ensuring that the benefits of progress are shared by all, not just a privileged few.

Imagine a future where clean energy is universally accessible, where education equips people to navigate complexity and solve global challenges, and where policies prioritize long-term well-being over short-term gains. This is not an unattainable dream but a realistic goal if we commit to working together.

Innovation with Purpose

Innovation is a driving force for progress, but it must be guided by purpose. The most impactful innovations solve problems while respecting the interconnectedness of life. They are designed not just for efficiency but for equity and sustainability.

The Switching Battery AC System is a case in point. Its design prioritizes affordability, accessibility, and environmental impact, ensuring that it benefits both people and the planet. This type of purposeful innovation sets the standard for how technology should evolve—not as an end in itself but as a means to improve lives and systems.

Educating for Interconnectedness

The future depends on how well we prepare the next generation to understand and navigate an interconnected world. Education systems must teach not only technical skills but also empathy, systems thinking, and collaboration. These are the tools needed to address the complexities of global challenges and to build solutions that are inclusive and sustainable.

Education must also emphasize the value of diversity. Different perspectives, disciplines, and cultures bring unique insights that enhance problem-solving and creativity. By fostering inclusive learning environments, we create leaders who can navigate interconnected systems with wisdom and compassion.

The Role of Leadership

Leadership in an interconnected future is about guiding, not dictating. It involves inspiring collective action, fostering trust, and aligning efforts toward shared goals. Effective leaders understand the balance between progress and preservation, the importance of reciprocity, and the power of collaboration. Whether leading a team, a community, or a nation, leaders must embody the principles discussed throughout this book. They must act with empathy, integrity, and vision, creating environments where everyone can contribute to and benefit from progress.

A Call to Action

The future we create together depends on the choices we make today. This is both a challenge and an opportunity. By embracing interconnectedness, prioritizing sustainability, and committing to equity, we can build systems that uplift everyone. Each decision—whether to innovate, collaborate, or educate—moves us closer to this vision.

This is not a task for governments or businesses alone; it is a responsibility shared by all of us. Whether through the technologies we adopt, the policies we support, or the ways we interact with each other, every action contributes to the future we create.

Building the Future

The principles of interconnectedness, reciprocity, and simplicity are not just ideas; they are tools for shaping a better world. They remind us that we are not isolated individuals but part of a larger whole. By aligning our actions with these principles, we can create a future that reflects the unity, resilience, and beauty of life.

The future is ours to build. It is a journey of collaboration, innovation, and shared purpose. Together, we can create a world where progress and preservation go hand in hand, where abundance is shared, and where humanity thrives as one interconnected community. This is the future we create together.

Chapter 21
Reflective and Proactive Awareness

Awareness is the lens through which we observe the unfolding of time and the arrangement of events within it. It is neither static nor singular but a dynamic and multifaceted process. To be aware is to actively engage with the frames of existence, understanding how they shape our perceptions, choices, and ultimately, our reality.

Reflective Awareness

Reflective awareness is the act of turning back the clock, not to dwell in nostalgia or regret but to seek clarity and meaning. When we reflect on past events, we are not simply recounting occurrences; we are reframing them in the context of what we know now. Reflection allows us to connect disparate events into a coherent narrative, offering insights that were perhaps invisible when the events first transpired.

For instance, consider the trajectory of a stream carving its way through rock. At any given moment, the stream's motion might appear random. But when viewed over time, patterns emerge—the stream's persistence, its adaptability, and the paths it has forged. Similarly, our reflective awareness transforms the raw material of experience into wisdom. It enables us to identify the recurring themes in our lives and recognize the lessons embedded within them.

Proactive Awareness

While reflection looks backward, proactive awareness gazes forward, framing possibilities before they manifest. It is the practice of envisioning outcomes, crafting strategies, and anticipating the interplay of forces yet to come. Proactive awareness is not prediction; it is preparation. It acknowledges the uncertainty of the future while positioning us to engage with it effectively.

Imagine an artist before a blank canvas. The act of painting requires an awareness of both the brush in hand and the vision of the artwork to be created. Proactive awareness is this dance between the present and the possible—balancing what is with what could be.

The Interplay of Reflection and Proactivity

True awareness exists at the intersection of reflection and proactivity. These are not opposing forces but complementary aspects of a unified whole. Reflection grounds us in our story; proactivity propels us toward its next chapter. Together, they create a cycle of learning, adaptation, and growth.

When faced with a critical decision, for example, awareness guides us to assess past patterns while imagining future implications. It's as though we are navigating a river, drawing on our memory of upstream currents to predict how the waters will flow downstream. By integrating reflection and proactivity, we are better equipped to choose paths that align with our values and aspirations.

Awareness as Presence

Beyond the reflective and the proactive, awareness is also about presence. To be fully aware is to inhabit the present moment with clarity and openness. It is to observe without judgment, to listen without the need to respond, and to exist without the compulsion to act. This form of awareness is often the most challenging, as it requires us to relinquish control and simply be.

Presence anchors us in the now, allowing us to connect deeply with ourselves and the world around us. It is in these moments of presence that we often experience the most profound insights, the kind that cannot be forced or fabricated but emerge naturally from a state of attunement.

Expanding the Frame

Awareness invites us to expand the frame through which we view life. Just as a photographer adjusts the lens to capture a wider perspective, we can widen our awareness to encompass not only our personal experiences but also the interconnectedness of all things. This broader awareness fosters empathy, resilience, and a sense of purpose.

When we observe frames of events over time—whether past, future, or present—we begin to see the threads that connect them. We realize that our lives are not isolated moments but part of a larger tapestry, woven with the threads of relationships, environments, and universal truths. This realization deepens our understanding of ourselves and our place within the cosmos.

The Practice of Awareness

Cultivating awareness is both an art and a discipline. It requires patience, curiosity, and intentionality. Practices such as mindfulness, journaling, and meditation can help us refine our awareness, making it sharper and more expansive. The true measure of awareness however is not in the practices themselves but in how they influence our daily lives.

When we are aware, we move through the world with greater grace and integrity. We make decisions that reflect our highest values. We respond to challenges with equanimity rather than reactivity. And we engage with others from a place of compassion and understanding.

Awareness is the foundation of a well-lived life. It bridges the past and the future, grounding us in the present while guiding us toward what lies ahead. It transforms the mundane into the meaningful, the fragmented into the whole. To cultivate awareness is to unlock the full spectrum of human potential, allowing us to navigate the frames of time with wisdom, creativity, and love.

Chapter 22
Beyond Binary: Open to Possibilities

A conclusion is not an end but an invitation—a call to live with openness, curiosity, and the understanding that we never know as much as we think we do. In every decision, we operate with uncertainty. We often possess less than half the information necessary for truly conclusive judgment. But recognizing this limited knowledge is a powerful starting point: it compels us to listen, to communicate, and to seek common ground.

The Power of Uncertainty

When we admit we don't know everything, we become motivated to learn more—especially about others. This has been my greatest lesson that I have learned in living my life where at 62, I am only just beginning to see a complete pattern.

Every transaction, conversation, or relationship has two sides. Our perspective alone accounts for perhaps 50% (or less) of the equation. Success, whether in business, personal relationships, or even in a court of law, hinges on understanding the other side's viewpoint.

- To "win" in life, we must aim for at least 51%, and that extra 1% comes from recognizing our own blind spots and appreciating the perspectives of others.

- By seeking deeper insights into our neighbors, coworkers, or even adversaries, we unlock empathy, collaboration, and harmony.

Sharing Common Ground

Although we may differ in appearance, age, language, rituals, or behavior, we share far more in common than we often realize. We are united by our humanity, our dreams, and our desire to live with dignity. Above all, we share Mother Earth—our home and our responsibility.

When we focus on common ground instead of superficial differences, we transform potential conflict into cooperation, and hostility into friendship. In the most powerful sense, we "destroy" our enemies by making them our allies, simply by taking the first step to communicate.

Balance and Harmony: The Key to Unlocking Possibilities

Life often presents choices as opposites: success or failure, giving or taking, right or wrong. But why settle for a false dichotomy when there are a hundred possibilities that might better serve us?

To live a life of oneness is to recognize that balance and harmony are not about compromise but about integration. It is about finding synergy where opposing forces complement one another, creating outcomes that exceed the sum of their parts.

This requires a mindset of openness to possibilities (OTP). It invites us to:

- Recognize our entitlement to explore all options without limiting ourselves to extremes.

- Examine the full spectrum of choices, seeking those that align with our goals and are actionable.

- Shift from competition to collaboration, creating synergistic solutions that uplift everyone.

Balance is not static—it is dynamic. It evolves as we grow, adapt, and learn. By cultivating this balance, we align ourselves with the natural rhythms of life and open doors to opportunities that binary thinking would overlook.

Awareness: The Foundation of Possibility

Living with openness begins with awareness. Every choice we make—big or small—contributes to the larger system in which we exist. Awareness enables us to see these connections and to act with intention.

To be aware is to understand that:

- Every action has consequences, both immediate and far-reaching.

- The good we do for others ultimately benefits us, while harm perpetuates cycles of suffering.

- Each of us has the power to create change through mindful choices.

Awareness is the first step in shifting from reactive to intentional living. It empowers us to take responsibility for our role in shaping the world around us.

Ask And It Shall Be Given

The greatest failure of humanity is our collective inability to communicate effectively—yet this is also our greatest opportunity for growth – to give and take in a positive sum game where win-win matters. Open communication fosters mutual understanding and invites creative problem-solving. Each moment we spend learning something new about a family member, neighbor, or colleague is a step toward harmony.

- **Begin at home:** Learn more about the people closest to you—your family, friends, or coworkers. Start by asking directly what you may give and what you would like to receive.

- **Extend outward:** Engage with your community, recognizing that everyone's perspective adds a piece to the puzzle. Again, ask how you can assist them and candidly inform as to what you seek from them as there's nothing free in this Universe – all exchanges, big or small, are for mutual benefit – where the value varies by moments and you are in-charge of the valuation.

- **Embrace the global context:** We all share one Earth, making international collaboration essential to address challenges like climate change, resource distribution, and social justice. Be open – admit or deny any statement you make and debate by placing your thoughts to be challenged as each one of us inhabiting the planet has an equal share and no one person is an expert of what is right and what is wrong.

Practical Ways to Embrace Possibilities

1. **Start with Yourself**

Change begins with individual actions that align with the principles of balance and harmony.

- **Simplify Your Life:** Focus on what truly matters and reduce unnecessary complexity.

- **Practice Reciprocity:** Give back to the systems that sustain you, whether through conservation, community support, or mentorship.

- **Reflect and Adjust:** Regularly evaluate the impact of your choices on others, the environment, and future generations.

2. Expand to Communities

The ripple effects of individual actions become transformative when extended to communities.

- **Foster Collaboration:** Work with neighbors, local businesses, and organizations to address shared challenges.

- **Educate and Empower:** Share knowledge about sustainability, equity, and systems thinking to inspire collective action.

- **Build Resilience:** Strengthen community support systems, renewable energy initiatives, and local economies.

3. Lead with Openness

Leadership rooted in openness and balance creates lasting change.

- **Model Balance:** Demonstrate harmony in decision-making, showing how to integrate diverse perspectives.

- **Listen Actively:** Understand the needs of others to build inclusive and effective solutions.

- **Encourage Innovation:** Support creative approaches that address complex problems for the collective good.

4. Think Globally

On a global scale, openness to possibilities calls for policies and innovations that reflect interconnectedness.

- **Support Equitable Systems:** Advocate for policies that address climate change, resource distribution, and human rights.

- **Use Technology for Good:** Promote innovations that democratize access to essential resources, like the Switching Battery AC System, which bridges gaps between energy sources.

- **Celebrate Diversity:** Embrace diverse cultures, perspectives, and disciplines as sources of strength and creativity.

What We Give, We Receive

At the heart of openness is the understanding that what we give to others ultimately returns to us. This is not just a spiritual truth but a practical reality. When we:

- Help a struggling neighbor, we strengthen the community and improve our own lives.

- Invest in sustainability, we secure a healthier planet for ourselves and future generations.

- Choose kindness over hostility, we foster environments where collaboration, not conflict, thrives.

This reciprocity is the essence of karma—the recognition that our actions shape not only the external world but also the internal one.

The Neighborhood Metaphor Revisited

As we reimagine living in a neighborhood where you reside in a luxurious home, while your neighbor's house is run-down. Your view from your window reflects your neighbor's struggles, not your wealth. But when you help uplift your neighbor—perhaps by contributing resources or opportunities—you improve the entire neighborhood, including your own quality of life. Property values rise, the environment becomes more vibrant, and both households thrive.

This metaphor illustrates a universal truth: helping others is not a sacrifice—it is an investment in mutual prosperity. It reflects the balance and harmony that comes from being open to possibilities.

A Life of Balance and Harmony

To be open to possibilities is to live with purpose and intention. It is to see yourself as part of a larger whole, to recognize the interconnectedness of all things, and to make choices that enhance harmony.

This way of living:

- Empowers us to explore options beyond binary extremes.

- Encourages us to prioritize collaboration and balance over competition.

- Reminds us that every connection matters, every choice counts, and every individual has the power to contribute to a better world.

Living with openness means understanding that balance is not an endpoint—it is a process. It is the ongoing practice of integrating diverse possibilities into harmonious solutions that serve both the individual and the collective.

Seeing the Whole Picture

The failure of mankind, resulting in wars and controversies, lies in its inability to communicate. We focus too much on what divides us—skin color, age, behavior, rituals, languages, and other superficial differences. The reality is that we share far more in common than what separates us.

At our core:

- We all love and seek connection.

- We all desire security and peace for ourselves and our families.

- We all share the same home—Mother Earth.

Appreciating these commonalities is far more productive than clinging to differences. When we begin to see the whole picture, we understand that our collective success lies in collaboration, not division.

Moving Forward Together

Being open to possibilities means accepting that life is neither black nor white but a rich tapestry of experiences and viewpoints. By acknowledging our limited knowledge, we invite others to fill in the gaps, to share their stories and insights, and to co-create solutions that serve everyone.

This final chapter is both a reflection on our journey and a pledge for the future:

- Openness reminds us that uncertainty can be a source of strength, pushing us to learn more.

- Communication binds us together, turning potential enemies into collaborators.

- Shared humanity is our anchor, urging us to focus on commonalities rather than differences.

In the grand scheme, the power of oneness is not just a concept; it is the fabric of our existence. By living with an open mind and heart, by actively seeking to understand rather than simply to be understood, we can tap into limitless possibilities for growth, harmony, and shared success.

Let us celebrate our differences by uniting around our shared planet, our collective future, and our desire for a world where every voice can be heard and every perspective valued. When we do so, we truly become architects of a brighter tomorrow—together.

The Power of Oneness

About Switching Battery

Switching Battery is an intelligent energy accumulator / harvester unit, built on Node Fusion Technology (NFT) as hardware and XY Logic as software. Together, it works like modifying a battery's DNA. Instead of mechanically linking entire batteries in one rigid setup, NFT lets you connect individual nodes electronically. By rearranging these smaller building blocks, the Switching Battery can seamlessly create AC or VFD waveforms—much like how tweaking genetic code can change an organism's traits.

Key Features:

AC/DC
DUAL CAPABILITY

- 24V Solar DC Input
- 110VAC Output

98%
CONVERSION EFFICIENCY

- 70% Software
- XY Logic Switching

UPS
UNINTERRUPTED POWER SUPPLY

- Flow- Thru Battery
- Simultaneous Charge/ Discharge

SAFE
LOW EMI FREQUENCY

- Below 5kHz
- Less Wear & Tear

EFFICIENT
SUSTAINABLE/ EFFICIENT

- Solar Integrated
- Fast Charging

*SDG 7
RELIABLE/ AFFORDABLE

- Low Cost
- Low Maintenance
- Lasts Longer

US PATENTS:
US11,398,735B2, US11,799,301B2, US9,193,266B2, US9,944,187B2, US2021/0218251A1, US2022/0368139A1, US2024/0047980A1, US2024/0170978A1, US2024/0186819A1

* United Nations' Sustainable Development Goal (SDG)

131

About the Author

Kannappan Chettiar (KC) is a visionary leader whose journey transcends disciplines, industries, and borders. From his entrepreneurial beginnings to his groundbreaking innovations in renewable energy, KC has consistently transformed challenges into opportunities, inspiring others to reimagine what's possible.

KC's academic journey reflects his unrelenting pursuit of knowledge and interdisciplinary excellence. He holds four law qualifications, including a Master's in Law specializing in Energy and Clean Technology from the prestigious University of California, Berkeley, and a degree in Finance and Economics from Michigan State University. He is a Fellow of the Singapore Institute of Arbitrators and an Honorary Fellow of the Chartered Institute of Marketing, underscoring his dedication to mastering diverse fields.

At the age of 27, KC demonstrated his entrepreneurial spirit by building a thriving private education business in Singapore, attracting students from across Asia. This early success was a testament to his ability to identify unmet needs and deliver impactful solutions. However, the devastating 2004 tsunami marked a profound turning point in his life.

Witnessing the fragility of the planet, KC felt a deep calling to address the challenges of climate change, shifting his focus to renewable energy and embarking on an ambitious journey to learn science and engineering after his retirement.

As the Founder and CEO of Switching Battery Inc., KC has redefined energy storage. His innovative approach challenges conventional methods, including Alessandro Volta's traditional battery connection techniques, by introducing Node Fusion Technology and XY Logic Switching. These advancements are detailed in his book *Switching Battery Para-SeriesConnections*. With nine US patents filed and five granted, KC's contributions stand as a testament to his ingenuity and relentless drive.

KC's work has earned global recognition, including being named Runner-Up at the IEEE Smart Grid Competition 2023, where he collaborated with international academic teams to advance energy solutions. His ability to unite diverse perspectives and foster synergistic growth reflects his personal philosophy: "KC" stands for both Knowledge and Collaboration—a belief in the power of collective intelligence to achieve shared success.

Through his innovations, teachings, and writings, KC embodies the spirit of progress, proving that with vision and determination, the boundaries of what is possible can always be expanded.

KC inspires others to dream bigger, act bolder, and harness the limitless power of human ingenuity to create a better, more sustainable world.

www.ingramcontent.com/pod-product-compliance
Lightning Source LLC
Chambersburg PA
CBHW050244170426
43202CB00015B/2911